Henry Granger Piffard

The Status of the Medical Profession in the State of New

York

Henry Granger Piffard

The Status of the Medical Profession in the State of New York

ISBN/EAN: 9783744670272

Printed in Europe, USA, Canada, Australia, Japan

Cover: Foto ©berggeist007 / pixelio.de

More available books at **www.hansebooks.com**

THE STATUS

OF

THE MEDICAL PROFESSION

IN THE

STATE OF NEW YORK.

BY

HENRY G. PIFFARD, M. D.

NEW YORK:
D. APPLETON AND COMPANY,
1, 3, AND 5 BOND STREET.
1883.

THE MEDICAL PROFESSION

IN THE

STATE OF NEW YORK.

FIRST ARTICLE.

From the New York Medical Journal for April 14, 1883.

COMPARATIVELY few are acquainted with the history of the events that led to the movement which resulted in the formulation of the New Code of Ethics of the Medical Society of the State of New York. The writer of this considers himself fairly well informed on the subject, and will give what appears to him the facts pertaining to the subject. In doing this, however, he must be pardoned for certain apparent digressions, since, in order that the matter may be correctly understood, it will be necessary to go back to the times that precede the promulgation of the "Old Code."

In New York the profession first became organized as a corporate body in the year 1806, and seventeen years later thought fit to lay down a set of rules for the government of its members. This action was deemed necessary in order to control some who appeared to regard medicine in the light of a trade rather than a profession, and who were regarded by their stricter brethren as medical freebooters rather than physicians. The result was the enactment in 1823 of the "System of Ethics of the Medical Society of the State of New York." This "system" or code inculcated two species of obligation, namely: those which the profession should bear to the public, and those which its individual members should hold to each other. This code was founded on and was an adaptation to local needs of an English work known as "Percival's Ethics." At this time there was little cause for uneasiness on the part of those who with propriety might be called "regular physicians," namely: those who were graduates of medical schools, and those who, after strict examination, were "licensed" to practice the profession by the bodies having due authority thus to license—to wit, the county societies. During these years, however, an irregular sect had come up, outside the profession, and who were commonly spoken of as the "steam-doctors" and "herb-doctors." These were men of no medical acquirements, and of varying degrees of honesty, who had embraced the doctrines of one Samuel Thomson, hailing from New England. Their chief therapeutic reliance was on vigorous sweats with the free use of lobelia and the utter condemnation of mineral and certain other powerful drugs. The clamor that they raised against the heroic treatment then practiced by the mass of the profession resulted in a most bitter feud, in which the laity, as is usually the case, took an active interest. The State government was appealed to, and for nearly twenty years the strife was kept up, sometimes the profession and sometimes the quacks being ahead. In 1827, however, the profession gained a definite and substantial victory, the medical act of that year placing in their hands the supreme control and regulation of the practice of medicine in this State. At this time the State society had less power over the county societies than at present, and the suppression of quackery was virtually left in the hands of the county societies, each having jurisdiction in its own district. The war against the irregulars, just mentioned, was kept up with more or less vigor in different localities. During the fourth decade of this century, however, a new form of irregularity appeared. I refer to the introduction of Hahnemannism or Homœopathy; terms which in those days were synonymous. This new form of heresy developed, not among the irregulars, but in the bosom of the profession itself. The adherents and advocates of the new doctrines were members in good standing of the county societies, and their brethren were unable to invoke the aid of the law to compel them to practice in accordance with the views and wishes of the majority. Another weapon, however, was brought into play, namely : social and professional ostracism. The public, as before, became interested in the quarrel. Many of the laity regarded the action of the majority as a species of oppression, and, as often happens, became partisans of the weaker party. During this decade the number of professed homœopaths increased and their adherents and supporters multiplied. The heretics were still members of the county societies, and there was no easy way of ridding the societies of them—that is, against their will. At that time the only way in which a member could be expelled from a society, and prevented from continuing his practice, was through a direct application to the courts. The courts, however, were unable or unwilling to give the societies the desired relief, feeling, perhaps, that they had no more right to interfere in matters of professional than of religious heresy. The societies, nevertheless, possessed one valuable franchise : They could prevent any new comer from practicing in their respective districts if they saw fit to do so. This afforded them the means, as they thought, of preventing the increase of homœopathy by accessions from abroad. About the year 1842 the Orange County society, I believe, availed itself of this power; and forbade a physician of homœopathic tendencies from practicing in that county. Fearing that he would in like manner be prevented from practicing in the other counties of the State, he gathered his friends together and, without much difficulty, procured the passage in 1844 of a law that deprived the county societies of their powers in this respect. This law, moreover, went much further than this, as it repealed the penal clause of the act of 1827 and virtually permitted any who chose, whether educated or not, to practice medicine in this State. This permitted quacks of all sorts and descriptions to ply their vocation without fear of molestation. This condition of affairs was maintained for thirty years, and there can be little doubt that this was the direct result of the injudicious action of the Orange County society, indorsed as it was by the then general sentiment of the profession throughout the State. Homœopathy now had free scope to extend its influence, and, as the evils of sectarian medicine were most keenly felt in New York and Pennsylvania, these States were among the foremost to consider how they might be averted. The result of this consideration was the birth of the American Medical Association. It seemed probable to this association that the most effective blow would be given to the new-born heresy, if the profession as a whole combined against it. It seemed necessary that the homœopaths as a body should be absolutely excommunicated from professional recognition and intercourse, and that the public at large should know it. In the code of ethics, and especially in the " consultation " clause, this sentiment crystallized. It was thought that the public, knowing that consultations were forbidden, would be afraid to intrust serious cases to the care of a homœopath who might be scores of miles distant from a colleague with whom he might consult. This action was, to say the least, exceedingly unwise as judged from a purely medico-political standpoint. In those days the chief therapeutic reliances of the profession were bleeding, purging, puking, blisters, and salivation. In contrast to this the homœopath offered medication that was not unpleasant to take, nor, apparently, disturbing in its effects. Is it a wonder, then, that many persons, finding themselves but trivially affected and yet desiring professional advice, preferred the milder to the severer medication? Happily recovering, they felt emboldened to trust even severer cases to the homœopath. The general profession, however, were blind to the teachings of these every-day occurrences, and it was not until Andral demonstrated in the hospitals of Paris that no treatment was preferable in certain diseases to the methods in vogue, that medical men awakened to the fact that in many cases they were doing their patients harm rather than good.

In England, Sir John Forbes learned the lesson, and endeavored to teach it to his countrymen. The reward he reaped was the scorn and hatred of his peers, and, after his death, the virtual adoption of his views (expectant treatment) by a succeeding generation. He simply taught that entire absence of treatment was often better than the heroic methods practiced by his colleagues. During these years the homœopaths, despite the opposition of the profession, increased in numbers and in influence, and, excluded by the "code" from joining the existing medical corporations, they applied to the State for authority to form corporations of their own. This they secured, with powers co-extensive and identical with those possessed by the older societies. Most of the older homœopaths joined the new organizations, but there was still left a certain leaven of unrighteousness, which the majority desired to get rid of. This could hardly be accomplished under existing laws, as the societies had not the power to prune their membership, except through an application to the courts. This was felt to be an inconvenience, and the Legislature was applied to for relief. Through the exertions of Dr. Oliver White and others, a law was enacted, in 1866, which greatly enlarged the powers of the county societies in this respect. The law in question permitted them to frame by-laws (subject to the supervision of the State society) which would enable them to visit expulsion on any member who should be guilty of irregular practices. The term "irregular practices" was a little indefinite, but was commonly understood to include employing remedies or methods that in any way resembled or savored of homœopathy. In the year following the passage of the act, the Westchester Medical Society invoked its aid to enable them to get rid of an obnoxious member who was charged with "irregular practice of medicine." Apparently the gravest charge against the member was the admitted fact "that he has purchased globules of sugar of milk by the pound from the Homœopathic Pharmacy in New York City," and "that he used these homœopathic globules in his practice, to induce his children patients to take the medicine which he prescribed for them." On these charges he was expelled by the county society, and, on the member's appeal to the State society, the action of the county society was sustained. The expelled member, if he desired professional affiliation, was now forced to join the homœopathic society. In this way the ranks of that body obtained occasional recruits. I do not mean that there were many formal prosecutions for the crime of giving the children a little candy, but the social and professional pressure was so great that many left the regular societies voluntarily, in order that they might obtain a little peace from persecution and be enabled to practice as they thought best. It is a curious fact that, while the regular societies *excluded* the use of certain medicines and modes of employing them, the homœopathic societies were really more liberal in this respect, none of them, I believe, having formally adopted the exclusive tenets of Hahnemann, or declared that their members must practice exclusively in accordance with the doctrine of similars. So far as I am aware, they never expelled any of their members who found that "confectionery" ("*Zuckerwaaren*," as the decision of a German court of justice recently termed it) was not always sufficient, and who supplemented it with a good dose of quinine or calomel. The homœopaths, then, were, practically at least, less exclusive than their elder brethren. Many physicians, who, led by a spirit of inquiry, investigated the homœopathic system, found some apparently striking verifications of the doctrine of similars, forthwith fancied that in these consisted the whole science and art of medicine, and made use of them as occasion required. This led to their ethical condemnation, and forced them into the established homœopathic organizations, in some instances long before they had any settled convictions on the subject. There is little doubt that the general effect of the "code" was, in many ways, to build up and strengthen the sectarian societies, not only by forcing men into them, but by exciting public sympathy in their favor, and thus aiding them politically. A house divided against itself can not stand, and a medical profession divided into hostile camps can not long retain the respect of the public, nor the good-will and assistance of the legislators.

Going back some years, we witness the birth of still another medico-political organization. The old herb- and steam-doctors, some of whom had picked up a smattering of medical knowledge, began to form voluntary organizations for mutual protection. Before long, they too aspired to corporate powers and governmental recognition. The existing feud in the profession rendered this a comparatively easy matter. Under the title of "Eclectics" they secured the same chartered rights as the other societies. There is little doubt that at this time the homœopaths aided the eclectics, believing that, by forming an alliance with them, they could prevent the regulars from in any way curtailing the corporate powers of either body. For obvious reasons an alliance of this sort could not be very long maintained. The homœopaths, as a rule, were educated men, while the vast majority of the eclectics were not. There was, however, another cause that tended to isolate the eclectics from the educated profession, and this was their code of ethics. Following the example of the regulars, the homœopaths adopted a code that was a verbatim copy of the American code, with the single exception of the consultation clause. The eclectics, however, adopted a code which in every important respect was the exact reverse of the American code. In their code they stated that it was proper for medical men to hold patents on surgical instruments, to advertise in the papers, to invite laymen to operations, to practice with secret nostrums, etc. Such practices most medical men regard as eminently improper, but this did not alter the fact that these men were in the eye of the law fully as regular as the very elect. This position they never would have obtained had it not been for the existing feud between the regulars and the homœopaths.

We now had in the State of New York three medico-political bodies, each with co-ordinate powers and co-ordinate jurisdiction. The differences between them were essentially as follows: The first or older organization and the second one were at variance simply on the question of practical therapeutics, while on questions of general medical polity they thought alike. The third organization differed from the others, both on the question of therapeutics and medical polity. One might suppose that the State of New York was by this time sufficiently afflicted, but such was not the case. A few years later, still another body, claiming to possess certain special therapeutic advantages, obtained corporate privileges and governmental recognition. But this was not all; the repeal of the penal clause of the act of 1827 permitted quacks and charlatans of every kind to come to the State and deceive the unwary in any manner that they chose. Such was the condition of affairs up to 1874. At this time some one—I have never been able to ascertain who—introduced into the Legislature a bill to regulate the practice of medicine in this State. This bill would appear on its surface to have been a desirable measure, but a careful study of it ought, at the time, to have revealed its true inwardness. If this bill was not originally drawn by the eclectics, it was unquestionably manipulated by them during its passage through the Legislature, and practically it turned over to them the licensing of every quack in the State who thought it worth while to pay them an examination fee of ten dollars. The majority of them, however, did not take even this trouble. They had enjoyed immunity for thirty years, and were not afraid to take the risks a little longer. As a matter of fact, the few prosecutions that were undertaken came to a lame and impotent conclusion.

This, then, was the state of medical affairs in New York about the year 1876. There were, first, the regular profession, enjoying chartered rights that dated back for seventy years, and consisting of men who were graduates in medicine, or licentiates (after examination) of the county societies; second, a sectarian offshoot from them, who were likewise educated men; third, a sect growing up by itself, and slipping into corporate existence while the first and second were quarreling—an exceedingly small number of these men had received a medical education; fourth, a sect that died almost at its birth; and, fifth, the horde of miscellaneous quacks who settled in the State during the times when this could be done with impunity. All of these things were brought about during the period that the profession were under the guidance of the code of ethics of the American Medical Association. Surely, if the object of this code were the suppression of quackery, its success can hardly be described as brilliant. How it is in other States, the profession there resident are the best judges. In what precedes and follows, I am speaking only concerning the State of New York. With quackery rampant to a degree never before witnessed in this locality, the problem to be solved was, What were the causes and what was the remedy? These questions could certainly not be answered off-hand and without consideration. The problem was one that required careful and earnest study, if a correct solution was to be reached. This study was undertaken by the then officers and censors of the Medical Society of the County of New York, whom the law had constituted the guardians and protectors of professional honor and professional interests within their jurisdiction. Almost the first conclusion at which this body arrived was, that the laws regulating the practice of medicine were palpably defective. With practically no law from 1844 to 1874, and after that a worse than no law, it was clear that, until an efficient statute was enacted, it would be impossible to expect much, if any, improvement in the affairs of the profession. At this juncture a gentleman who had acted as the legal adviser of the society offered to prepare a suitable bill, and endeavor to procure its enactment. His offer was accepted. A bill was prepared, and introduced in the Senate. This bill was referred to a committee, and an hour was assigned for its consideration. At the appointed time the advocates and opponents of the bill presented their views to the committee, which was represented solely by its chairman, Dr. Ray V. Pierce, of Buffalo, a noted medical advertiser, and a member, we believe, of the eclectic organization. The bill did not meet the approbation of the committee, was not reported favorably to the Senate, and did not become a law. The next attempt to secure suitable medical legislation was made by the State Medical Society. In 1880 it instructed its Committee on Legislation to prepare a proper law and submit it to the Legislature. This was done, and the Medical Act of 1880 was the result. Prior to the passage of this act there were in the State upward of one hundred and fifty bodies that were competent to legitimize practitioners of medicine. The act in question reduced the number to thirteen. Of these bodies, two were eclectic, two were homœopathic, one was nondescript, and the rest pertained to the regular school. Within the past year two of these bodies have been declared illegally constituted, and their career has ended. In 1882 an attempt was made by the State society to reduce the number of licensing bodies to one. The bill which was drawn for the purpose of effecting this object did not become a law. The law of 1880 remains in force, and under it the entire responsibility in regard to the licensing and legalizing of practitioners in the State rests with the medical colleges of the State, while prosecutions for violation of the law may be undertaken either by individuals or the county societies. In New York County these prosecutions have been numerous, and usually successful. Thus far but one flaw or serious imperfection in the law has been discovered—namely, that the penalty for perjury in connection with registration is not sufficiently severe. Shortly after its adoption, our Pennsylvania brethren procured the enactment in that State of a law identical in its main features with the New York law. That the New York law is all that is to be desired, or that it is the best medical act in this country, is far from being claimed. In fact, I believe that Illinois and North Carolina have better ones, both from a theoretical and practical standpoint.

From this *résumé* of the medico-political situation it will be seen that, after a sharp fight with quackery, the profession obtained the upper hand in 1827 : that for several years it retained this control ; that in 1844 it lost its power, and failed to regain any of it until 1880 ; that even now it does not possess the full powers and privileges that it formerly enjoyed.

SECOND ARTICLE.'

From the New York Medical Journal for April 28, 1883.

In our last we presented the medico-political or legal relations of the profession. In this we will consider the medico-educational.

The oldest of the existing medical colleges is the College of Physicians and Surgeons. This institution was chartered, not directly by the State, but by the regents of the university, in the year 1807. It was at that time virtually the same corporation as the New York County Society, or, in other words, the county society was constituted a medical faculty, with authority to teach and grant diplomas. The intimate relationship was not long maintained. The teaching body obtained independent powers, and was subservient, to a slight degree only, to the county society as such. The relationship, however, was not wholly dissolved, for, a few years later, when the college exhibited an unbecoming laxity in the granting of degrees, the society exercised its powers and influence to break up these practices. Since that time there has been, so far as we are aware, no special scandal connected with the management of its affairs. As an alumnus of the institution, we feel pride in stating that it has, before all others, been careful in the exercise of its corporate powers. This statement, however, is not true of some other colleges that were in existence during the early part of this century. The granting of diplomas was so lax that the State declared they should no longer be a license to practice ["The degree of doctor of medicine conferred by any college in this State shall not be a license to practice physic or surgery," Act of 1827, Sec. 21]. Subsequent to this time several new colleges were organized, which, in their charters, obtained the right to make their diplomas licenses to practice. At the present time the diploma of every legally incorporated medical college in the State carries with it the license to practice. On the other hand, no medical college in this country or elsewhere issues a diploma which entitles its bearer to practice in this State, except with the approbation of one of the college faculties of this State. It matters not whether the candidate has drawn his inspiration from Gross or Buchanan, he must first satisfy a college faculty of this State of his fitness to practice before he can become a legally qualified practitioner in this State. As before noted, the entire responsibility concerning the qualifications of practitioners coming into the State from without the State rests with the colleges. For the assumption of this responsibility they are entitled to exact a fee of twenty dollars in each case. Until recently there were thirteen colleges capable of exercising these powers. Of these, eight professed to teach non-sectarian medicine, located, four in New York city, one in Albany, one in Syracuse, one in Buffalo, and one in Brooklyn. Two taught homœopathy, both located in New York ; two were of the eclectic persuasion, both in New York ; and one, the " College of Physicians and Surgeons " of Buffalo, was a nondescript. These colleges all possess the power of granting the degree of doctor of medicine, and their diploma carries with it the license to practice in the State, after the graduate shall have complied with the registration law of 1880. The legal requirements for graduation are the same in all—namely, three years' pupilage with a legally qualified practitioner (not necessarily of this State), attendance on two full courses of lectures, the last of which in the college granting the degree, and the passage of a satisfactory examination in the seven principal branches of medical science. It is safe to say that the requirements in the matter of examination have not been identical in the thirteen institutions. As regards the regular colleges, there have been no public scandals connected with improper graduation of candidates, at least of late years. The same can be said of the homœopathic colleges, but can not be said of either of the eclectic colleges. The " Eclectic Medical College " of New York has been very strongly suspected of issuing diplomas contrary to law. Suspicion was first directed toward the other eclectic institution, known as " The United States Medical College," in consequence of the receipt by the officers of the New York County Society of a communication from the Illinois authorities asking the status of said college. The communication further stated that a person armed with a diploma of that institution had applied for a license to practice in Illinois, under circumstances that led them to suspect that he had obtained his diploma illegally. This led the officers of the society to watch the college, and, on examination, they became satisfied that the college itself was not legally incorporated, and they instituted a suit against it. The Supreme Court of the State has, within the past few weeks, rendered a decision to the effect that the college was *not* legally incorporated, and hence that none of its diplomas are legal. In Erie County the same may be said. The College of Physicians and Surgeons of Buffalo stood on exactly the same footing as the United States College, and a similar suit against them has resulted in a similar decision from the Supreme Court. We have now but eleven medical colleges, against thirteen of a year ago. How much further the shrinking process will extend it is impossible to foresee. " The mills of God grind slowly, yet they grind exceeding small," and the profession of this county may rest assured that, if they give their officers proper moral and financial support, illegal practice and quackery of all sorts will be an exceedingly hazardous pursuit.

It may, we think, be truthfully stated that, at the present time, quack-cry * and unqualified practice prevail here to a less extent than in any other State in the Union, with the exception of the States of Illinois and North Carolina. On the other hand, the States in which it flourishes most luxuriantly are Massachusetts and Pennsylvania, the latter State claiming to be the banner State of the old code, while the former has a special code of its own that is, in some respects, even more stringent than that of the American Medical Association. If we turn now to the States of Illinois and North Carolina we find that in the former quackery flourished to an alarming extent just so long as the profession was actively aggressive toward sectarian medicine. As soon, however, as it joined hands with sectarianism for the purpose of putting down quackery, it then began to triumph over the common enemy. This joining of hands occurred when the Illinois State Board of Health was established, in which were representatives of the regular, the homœopathic, and the eclectic schools. In North Carolina the case was somewhat different. In that State the profession had never allowed the subject of sectarianism to trouble them very much. If a homœopath by any chance settled among them, they very sensibly let him alone. They neither persecuted nor prosecuted him. They gave him no opportunity to play the martyr, or to parade his grievances in public. We have been curious to learn the outcome of this policy, and, on inquiry, have been informed that, of fourteen hundred physicians in that State, there are but six homœopaths, and, so far as known, no eclectics. In contrast to this let us cite the neighboring county of Kings in our own State. Many years ago a gentleman of homœopathic proclivities applied for admission into the county society. He was refused membership. He carried the matter to the courts, and obtained a decision in his favor. He did not join the Kings County Society, however, as in the mean time a homœopathic society had been formed of which he became a member. By the continuance of the same policy the Kings County regulars succeeded in building up against themselves a pretty strong sectarian organization, and now rejoice in one homœopath to about every six regulars, a larger proportion of homœopaths than will be found, we believe, in any other portion of the United States. New York city has about one homœopath to ten regulars.

This digression aside, we return to the subject immediately under consideration, namely, the medico-educational status of this State. Of the eleven medical colleges, three may be placed in the front rank as regards importance and facilities for medical instruction. Ranking with them are two colleges in Pennsylvania and one in Massachusetts. These six colleges compete for and obtain the patronage of the better class of students, the one in Massachusetts, however, possessing a higher standard for entrance than the others. One of the colleges of this State emulated the example of Harvard, and declared that it would require an examination of the student's

---

* By quackery we here mean the practice of medicine by uneducated and legally unqualified persons; while by sectarian medicine, practice in accordance with some special method or doctrine.

fitness before permitting him to matriculate. This declaration was regarded by the profession at large as an indication that the faculty of the college were determined to elevate the standard of medical education in the State, and in this way contribute to the elevation and maintain the dignity and honor of the profession. We all know how the experiment terminated. After one year's trial the faculty reconsidered its resolution to require a preliminary examination, and resumed its former status. The two other colleges, however, have made some substantial advances ; one of them, by enlarging its building, adding laboratories, etc., has increased its facilities for teaching, and the other has materially lengthened its lecture course. At the present time the clinical advantages, the facilities for instruction, and the quality of instruction actually given in this city, are, we believe, unsurpassed by any to be found elsewhere in this country. The other colleges in the State have, according to their opportunities, done well, and, in some respects, have shown a more progressive spirit than the metropolitan institutions.

This, then, is the medico-educational status at present. What it will be in the future it is impossible to foresee. There are evidences, however, that thoughtful minds in the profession are looking and hoping for still greater improvement. This may take the shape of a single board of examiners, or the establishment of a medical college so largely endowed that the number of the students and of graduates will not be a material factor in the requirements of the college, or possibly the State or the municipality may itself assume the prerogative, as in several European countries, of educating those who aspire to be physicians. These, however, are questions that do not appear to exact immediate settlement, nor is such settlement at present possible.

THIRD ARTICLE.

From the New York Medical Journal for May 5, 1883.

Having considered briefly, but we believe accurately, the medico-political and medico-educational status of the profession in this State, we will now take up the question of its medico-ethical position and requirements.

As already noted, the earliest attempt at ethical regulation in this State was the adoption of the System of Ethics of the Medical Society of the State of New York in the year 1823, long before any other State had thought it worth while to move in the matter. This code remained in force until 1880. About the year 1850 the "Code" of the American Medical Association was also adopted by the State society, but without the repeal of the older code. The profession of the State were, therefore, under the guidance and governance of two distinct codes, the respective provisions of which were not altogether in harmony. Thirty-five years ago there were those who preferred the old State "System" to the "Code" of the American Medical Association; but, as by the adoption of the American Medical Association code there was a prospect of national unity on . the matter, they yielded their preferences, and consented to be bound by both, thus accepting a measure of ethical responsibility in excess of that borne by the profession in any other State in the Union. At this time there was no other organized section of the profession to question the propriety of this code, or to propose the adoption of another. A few years later the homœopaths became organized, and acquired chartered privileges. They were, as already stated, an outcropping from the general profession, and thought best to follow its example and adopt a code. This code was an almost exact transcript of the American code, with the exception of the paragraph relating to the question of consultations. On this point they differed from the older code as follows:

"A complete medical education, of which the diploma of a medical college is the formal voucher, furnishes the only presumptive evidence of professional acquirements and abilities. But the annals of the profession contain the names of some who, not having the advantages of a complete medical education, became, nevertheless, through their own exertions and abilities, brilliant scholars and successful practitioners. A practitioner,.

therefore, whatever his credentials may be, who enjoys a good moral and professional standing in the community, should not be excluded from fellowship, nor his aid rejected, when it is desired by the patient in consultation. No difference in views on subjects of medical principles or practice should be allowed to influence a physician against consenting to a consultation with a fellow-practitioner. The very object of a consultation is to bring together those who may, perhaps, differ in their views of the disease and its appropriate treatment, in the hope that, from a comparison of different views, may be derived a just estimate of the disease and a successful course of treatment.

"No test of orthodoxy should be applied to limit the freedom of consultations. Medicine is a progressive science. Its history shows that what is heresy in one century may, and probably will, be orthodoxy in the next. No greater misfortune can befall the medical profession than the action of an [influential association or academy establishing a creed or standard of orthodoxy or regularity. It will be fatal to freedom and progress in opinion and practice. On the other hand, nothing will so stimulate the healthy growth of the profession, both in scientific strength and in the estimation of the public, as the universal and sincere adoption of a platform which shall recognize and guarantee:

"1. A truly fraternal good-will and fellowship among all who devote themselves to the care of the sick.

"2. A thorough and complete knowledge, however obtained, of all the direct and collateral branches of medical science, as it exists in all sects and schools of medicine—as the essential qualifications of a physician.

"3. Perfect freedom of opinion and practice, as the prerogative of the practitioner, who is the sole judge of what is the best mode of treatment in each case of sickness intrusted to his care."

The additional sections of the homœopathic code so closely resemble those of the American code that they need not be quoted. They recognize the impropriety of advertising, of patenting surgical instruments, of practicing with nostrums, keeping secret the nature and composition of medicines used, etc.

A few years later, when the eclectics became organized, they also adopted a code—one that was likewise based on the American code, and differing from it mainly by reversing the intent of many of the most important sections of the older code as follows:

"Article III (*Eclectic Code*).

"Medical men have an undoubted right to bring themselves and their claims before the public by every fair and honorable means, as much as any other class of men. They may enter into general or special practice as they may consider best adapted to their interests or to their peculiar views; they may introduce themselves to the notice of the public by printed cards or other publications, by public or private lectures, or by the publication of certificates of cures actually performed. The presence of laymen at operations is by no means objectionable if both patient and operator shall consent, as it tends to make the skill and ability of the operator better known in the community, etc.

"Article V.

" A medical man having invented any surgical instrument, or discovered any new or valuable medicine, it becomes his capital, and it is not unprofessional for him to obtain a patent for the same. . . . A physician may employ, in his own practice, a medicine or compound known only to himself; it is his capital, and there is no authority in the land which can compel him to divide that capital among others by disclosing his remedy, save his own benevolence and philanthropy," etc.

Such is the " Code " of a body of men who would never have received governmental recognition if the educated members of the profession had not been engaging in a bitter internecine warfare.

This was the status of nominal ethics until within a recent period. The regular profession and the two sectarian bodies each had its code of ethics, differing from the others in the manner that we have seen. While the American code held sway over the great mass of the profession, sectarianism was increasing in power and influence. An evil which at its birth could have been easily controlled by wise measures was, on the other hand, injudiciously stimulated to an abnormal growth. For many years the American code had, in great measure, lost its vitality, and its edicts were not respected. One form of impropriety after another came to the surface, which it appeared unable to rectify or control. Eminent members of the profession began to violate not only its spirit, but its letter, and the corporate bodies of which they were members appeared unwilling or unable to subject them to discipline. About fifteen years ago the New York Academy of Medicine did discipline one of its members for consulting with a homœopath. Their experience on that occasion led them, quite wisely, to refrain from a repetition of the experiment. A few more such attempts would undoubtedly have led to the disruption of the Academy, and, in all probability, to a forfeiture of its corporate privileges. The suspension of the late Dr. Gardner from his rights as a Fellow of the Academy undoubtedly acted as a partial restraint on the other members of that body, and more especially on those who were comparatively young in the profession, or without sufficient influence to shield them from prosecution. As a matter of fact, two Fellows of the Academy permitted it to be publicly understood that they consulted with homœopaths, and would continue to do so as often as they pleased. Despite this fact, these members have never been brought to the bar of the Academy for discipline.

In 1865 ethical affairs were in such a state in New York that the late Dr. Oliver White saw fit to send to the Comitia Minora of the county society a communication, from which I extract the following : " It is patent to us all, Mr. Chairman, and it is daily manifest, that members of our profession, once occupying honorable positions in it, have lost their standing among us by their own disreputable, dishonorable, and empirical practices, in violation of all medical ethics ; and, insomuch as we are not permitted by the laws of the State to discipline or expel unworthy members from the county medical societies, except through the courts ; and, insomuch as we deem it both just and proper that our county society should be the custodian of its own honor, and the conservator of its own morals—therefore do I earnestly entreat the Comitia Minora to draft a memorial to the Legislature of the State, praying that honorable body to grant the county medical societies throughout the State relief from the oppressive disabilities aforementioned ; and that the Comitia ask the approval of the society to the proposed action in this matter."

The result of this action was the passage of the Medical Act of 1866, which gave the county societies almost plenary powers in matters of discipline. At the meeting of the county society, held June 4, 1866, a resolution was offered to the effect " that a committee of three be appointed to examine our list of members, and report the names of those whose connection with the society should be dissolved, and also what steps should be taken to accomplish this result." This resolution was referred, with others, to a committee of five, which met and considered the matters referred to them. This latter committee reported at the September meeting of the society that they could not purge the roll of membership as proposed, and recommended that " no further action be taken upon the matter at present," and requested " to be excused from further deliberation." This was certainly a rather impotent conclusion of the effort to purify the morals of the profession. Since then very little has been done in the way of attempting to check violations of the code.

This brings us to the year 1876, when the first open and bold proposal to repeal this code was made by Dr. J. Marion Sims, in his presidential address before the American Medical Association. From this address we extract the following :

" Here common sense and common interests have silently, almost imperceptibly, established a higher law that overrides the code and leaves it inert."

" The code of ethics is violated every day, either willfully or ignorantly, not only by the rank and file, but by men high in the profession—men who are considered leaders, advanced thinkers, and workers."

The proposition of Dr. Sims to abolish the code produced a profound sensation. Many thoughtful persons asked themselves whether the code as it existed, but unenforced, was doing any good, while others asked whether or not it was not doing absolute harm. It may be safe to say that by the majority Dr. Sims's proposition was looked on with disfavor. At all events, no action was taken in support of it by the association at the time. This is not surprising when we consider the composition of this body, made up

2

as it is of representatives from all parts of the country, very few of whom had examined the subject with any care. The hatred of sectarianism was so great that men seemed unable to calmly consider how it could be abated. Nothing was done. Shortly after this the writer found himself, as an officer of the county society, face to face with questions in ethics that must be met. The code was violated daily, both by those of high and low degree ; but discipline was rarely asked for, except to gratify some personal malice. The two most obvious violations consisted in mixed consultations, and a striving after notoriety through the medium of the public press. These evils the officers of the society found themselves powerless to combat. This may appear to be a strange statement, but the facts are as follows : In reference to mixed consultations, the apparent spirit of the code was rendered nugatory by a change of base on the part of the homœopaths themselves. Their State society adopted a formal resolution,* in which they declared that for the future they would not adhere to the exclusive doctrines of Hahnemann, but would use such other methods as individually they saw fit. A careful comparison of this resolution with the consultation clause of the code makes it clear that the homœopaths had thus *technically* freed themselves from the ban, and that it would be impossible to discipline a member of the county society who should consult with them.

The second difficulty that embarrassed the Comitia was the matter of newspaper notoriety which certain members gained through "interviews," and through certificates given in favor of certain mineral waters, etc. This was a new form of impropriety, against which there was no provision in the by-laws of the society, and the Comitia were, therefore, powerless to take official cognizance of the matter. The president of the society, however, took on himself the burden of appealing personally to the offending members. Strange as it may appear, some of them referred him to the code of the American Medical Association, and claimed that their conduct was not only blameless, but praiseworthy. Curiously, an examination of this code appeared to support their claim. Under these circumstances, the Comitia had but one resource—namely, an appeal to the State society, in the hope that it would enact such laws as would enable the county societies to effectually deal with the evils referred to. Such an appeal was made to the State society, at its session in 1879. It was disregarded, and no relief

---

* *Resolved*, That, in common with other existing associations which have for their object investigations and other labors which may contribute to the promotion of medical science, we hereby declare that, although firmly believing the principle *similia similibus curantur* to constitute the best general guide in the selection of remedies, and fully intending to carry out this principle to the best of our ability, this belief does not debar us from recognizing and making use of any experience, and we shall exercise and defend the inviolable right of every educated physician to make practical use of any established principle in medical science, or of any therapeutic facts founded on experiments and verified by experience, so far as, in his individual judgment, they shall tend to promote the welfare of those under his professional care."—*Adopted by the Homœopathic State Society*, February, 1878.

was afforded by the State society. The Comitia, however, were not altogether discouraged, and requested the writer to correspond with the chairman of the proper committee of the State society and ask him to make a personal investigation of the matters in question. A lengthy correspondence resulted. It was early conceded that the American code did not afford protection against the rapidly increasing certificate nuisance, and, if this ought to be stopped, some new and more effective rule must be adopted. The matter of consultations, however, presented much greater difficulties in the way of settlement, as on it hinged the whole question of sectarianism and its influence for good or ill on the profession. The present writer's views on the subject were presented in the form of a letter under date of November 28, 1879. From this letter I shall now quote at some length:

"The question" (of sectarianism) "is a grave one, and demands serious examination at the hands of the enlightened members of the profession, and never more so than at the present time. The subject must be looked at in its several aspects, and regarded from the standpoints of medical politics, of doctrine, of utility, and of its present *raison d'être*.

"In respect to the medico-political aspect of the question, it may be stated that the existence of sectarianism in medicine is a great evil, perhaps the greatest that at present oppresses the profession, and tends to injure it in the esteem and respect of the public. The State has seen fit to recognize three kinds of practitioners, who are in the position of public antagonists, each claiming that the general methods of treatment pursued by them are superior to those employed by the others, and each decrying and speaking in derogatory terms of the others. The public is not competent to decide to which of these the greatest measure of merit pertains; and, in the majority of instances, the choice of a medical attendant is the result of considerations that need not be entered into at present. The public, however, interests itself to a greater or less degree in the controversial elements of the question, and the result is a certain distrust of and lack of confidence in all three.

"The venerable Hufeland, nearly fifty years ago, in a very able essay on the subject, pointed out the evils that would result from sectarianism in medicine, and anticipated the statement made in the last paragraph. He wrote: 'Nothing is, on the whole, more prejudicial to our art, nothing tends more to diminish public confidence in it, than *a public quarrel, and the public expression of a mutual depreciation of one another by its professors.* All who have the honor of the art at heart must lament such open bickerings, and do all they can to prevent them. The public is only too disposed to interest itself and to find amusement in them. Has it not already come to such a length that our dissensions are paraded on the stage, just as in the time of Molière? And do we not feel that just as the estimate of our art in general decreases, so every one, to whatever party he may belong, loses somewhat?' . . .

"If, now, it be granted that sectarianism is an evil, does it not behoove the profession to consult together as to the best means to abate it? Before, however, this question is answered, it will be necessary to consider whether the abatement of this evil would bring about others that were still greater. The correct solution of this necessitates an acquaintance with a large number of facts, *pro* and *con*, which must be duly weighed, and the probable effects of any change in the present status carefully estimated. The opinion that I have personally formed is, that a certain amount of temporary inconvenience would ensue, to be followed by advantages that would more than counterbalance it. . . .

"If the abolition of legalized sectarianism is desirable, the methods of its accomplishment must be considered. To this end but two are known to me—namely, force and persuasion. Force has been tried, and has failed as regards the homœopaths. The more severe the exclusive enactments against them, the more they have seemed to flourish. The American Medical Association, in 1847, by the enactment of the consultation clause in the code, thought, by thus throwing odium on them, that the people would sustain the profession and refuse to employ the homœopath. In this the association was mistaken. . . . The action of the association, therefore, by excluding the early homœopaths from professional intercourse, simply caused them to unite the closer among themselves, each befriending and defending the other in time of need, and all uniting for the promotion of certain common objects, more especially the acquisition of the confidence of the people, and the attainment of governmental recognition. The measure of success that has attended their efforts we are all witnesses of to-day. It will, therefore, be readily granted that the policy of the association has not accomplished its object, i. e., the suppression of homœopathy, and I seriously question whether a continuance of this policy will not prolong and aggravate the present evils.

"Before, however, any other method be attempted, it is expedient that we should be accurately acquainted with the political and doctrinal status of the homœopathy, not of 1847, but of to-day, and we should more particularly regard the matter in its relations to the people and the profession of the State of New York. The homœopaths of this State may be divided into two pretty sharply defined groups. One group holds that the proposition "similia" is of great service in the selection of drugs where these agencies are requisite in the treatment of disease. They respect Hahnemann as a prominent promulgator of this doctrine. They reject, however, his theory of dynamization, they reject his peculiar views regarding the origin of chronic diseases, they reject his views as to dosage, and disbelieve or deny his statements concerning the efficacy of infinitesimals.* The other group of homœopaths pretend to hold strictly to all of Hahnemann's

---

* "The homœopathy of to-day has also shaken from its feet the dust of more than one worthless theory. Although within its ranks are still numbered some so-called high dilutionists, its leaders have long ceased to insist upon infinitesimal dosage as an essential principle of treatment."—BEARD, "Popular Science Monthly," February, 1883.

doctrines, and consider themselves his only genuine followers. An aggressive movement on the part of this latter party led the liberals to a countermovement, which resulted in the adoption by their societies of the resolution we have given above, and a decided split in their ranks.

"The regular profession has now an opportunity of settling the homœopathic difficulty in a very simple manner. Let it be understood that it is willing to receive into fellowship those who have practically abandoned Hahnemann's homœopathy, on condition that they also abandon the name, calling themselves, and permitting themselves to be called, *physicians* simply. It is probable that during the first year or so but a small number would avail themselves of the opportunity of joining the county societies. Later they would come in more freely. This would result in a return to the State of the chartered rights now possessed by them, and the removal of sectarian and offensive titles from the hospitals, dispensaries, colleges, and journals now controlled by them."

The above are the views that were held by me at the date that the above letter was written, and are in substantial accord with those that I hold to-day.

There is little doubt that, if this course had been pursued at that time, when the homœopaths were in so badly demoralized a condition, to-day there would have been no organized body, in this county at least, occupying an antagonistic attitude. The writer's position was, and is, that social and professional absolution be accorded to all who are willing to renounce exclusivism and unite with the main body of the profession.

In the number of this Journal for April 7th, pages 372 and 373, Dr. Flint would appear to be willing to go even further, and accord professional recognition to all, whatever their belief or practice, provided only they discontinued their connection with sectarian societies.

FOURTH ARTICLE.

' From the New York Medical Journal for May 26, 1883.

At the annual meeting of the Medical Society of the County of New York, held October 25, 1880, its Committee on Ethics made the following report: *

. . . "Almost without an exception, therefore, the work of the committee has been confined to complaints against members of the society for public advertising, or methods that are regarded as at variance with the spirit, if not the letter, of that portion of the American Code of Ethics embraced in Chapter II, paragraphs three and four.

"The committee approached this well-known field of action, where an almost ineffectual skirmish has long been kept up, with feelings of great uncertainty as to what the result of their efforts would be, and with a desire to perform their unpleasant duty without giving unnecessary offense to any. The committee, although to some extent shielded from personal attacks by its official character, has been unable in all instances to perform its duties without censure from individuals with whom it has been in communication; and in other instances, where requirements were made under the committee's interpretation of the law, it has encountered firm opposition. Previous to this committee's appointment, efforts had been successful in securing the withdrawal of physicians' mineral-water testimonials from the public press, but it was well known to the committee that all attempts to suppress those still appearing in the medical journals had been in the main unavailing; indeed, these futile efforts were treated as an encroachment on the rights of those concerned.

"The society, when appealed to on this subject at a meeting held April 22, 1878, adopted a resolution clearly expressing its disapproval of the practice of giving certificates to be used in bringing to notice 'any drug, nostrum, mineral water, wine, or other proprietary article intended to be used as a medicine or remedy in disease, or to any patented instrument or appliance that is intended for medical or surgical use.'

"This resolution, which, morally at least, had all the force of a by-law of the society, was at the time of its adoption brought to the notice of every member of the society. The effect, however, was not what had been expected, and the commercial pages of the medical press still teemed with advertisements of trademarked preparations, etc., bearing the sanction of medical men. . . .

"The committee, after mature deliberation, being encouraged by the success

---

* I omit such portions of the report as have no bearing on the present question.

others had already attained, and fortified by the resolution of the society above alluded to, thought that it was clearly their duty to make another appeal to those who still considered that they had a right to give their sanction to the articles under consideration in a manner objectionable to the profession in general. The committee, therefore, adopted the following at a meeting held January 7, 1880: '*Resolved*, That, in view of the fact that these certificates are offensive to a majority of the profession, and that their continuance is an injury to professional tone, the Committee on Ethics respectfully requests the gentlemen concerned to take measures to have them discontinued.' This resolution, together with that adopted by the society, was printed in the form of a circular, and copies of it were sent to all whose names had been reported to the committee, with a request that they would signify to the committee what course they intended to pursue in the matter. The total number to whom the circular was sent did not, perhaps, exceed two dozen.

"As a result of this action, a very small number signified their intention to withdraw the objectionable testimonials, and the committee has been informed that they have done so. In a few instances, letters in vindication of certificate writing were received, but it is believed that the position assumed in defense is not tenable, for in the advertisement of lactopeptine, for instance, the virtues of the remedy are extolled in a manner rather to arrest the public eye than instruct the physician; and this of a preparation where the method of manufacture is kept secret, and where the copying of its name by any one would render him liable to prosecution. The indication of the constituents of this preparation does not relieve it from the objection held against trade-marked and proprietary articles.

"The greater number to whom the circular was sent, however, failed to respond to the committee's request, and their certificates continue to appear.

"It will thus be seen that the committee has advanced this work but little, for, so long as any member can permit of the publication of these certificates with impunity, the majesty of the codes of ethics is not maintained. And, now that the society has continued to experience defeat in this matter, it may be well, before entering the contest again, to inspect its position and strength. The committee has, therefore, made a careful examination of that portion of the American Code of Medical Ethics and of the System of Medical Ethics of the Medical Society of the State of New York bearing on this subject, and it seems to it that, although their provisions may have been sufficient for the time when they were adopted, a gradual transformation in the character of the abuses alluded to has taken place, and, instead of secret remedies, there has grown up the proprietary and trade-marked article, which requires the investment of a large amount of capital. Secrecy has ostensibly been removed as to the constituents of these goods, but their manufacture or imitation is successfully prevented by patents and trade-marks. That they owe their chief value to professional testimonials and skillful advertising may well be believed. The committee has failed to find anything in the codes referred to sufficiently explicit to give them plenary power to take further action. If, therefore, the society desires to prevent its members contributing to trade interests in the manner above alluded to, and thus injuring its own, it has ample power under the State statutory laws to make the resolution of April 22, 1878, a by-law of the society. The experience of the committee leads it to believe that no other course will accomplish the end desired. . . .

"In this connection, it may be pardonable for the committee to state that its experience during the past year has, in a forcible manner, demonstrated the inadequacy of the present codes of medical ethics to the existing demands of the profession. The code adopted by the American Medical Association thirty odd years ago has in many respects become obsolete; what were deemed offenses then are no longer regarded in the same light. *Per contra*, the ingenuity of man has developed practices which were unknown when the codes, national and State, were established, and hence were unprovided for. The code of the American Medical Association contains a mass of sentimental advice which, together with its moral platitudes and verbiage, would seem to suggest the necessity for its revision. Our own System of Medical Ethics, which the State society adopted in 1823, and which has since been subjected to but few alterations, is, perhaps, even more obsolete than the code above alluded to. The profession is now in no sense guided by these codes; nor does it seem desirable to hold it together either by the regulations that pertain to trades-unions, or by the moral platitudes of existing codes, but it rather requires for its wholesome government clear and business-like regulations, backed up by our ample statutory laws, leaving the matter of moral maxims and precepts, as well as personal manners, to the social conditions that surround the individual."

This report was signed by Drs. Samuel Sexton (Chairman), James R. Leaming, W. M. Polk, J. D. Bryant, and Clement Cleveland (Secretary).

The foregoing was printed, and distributed to the members of the society.

At the annual meeting of the Medical Society of the State of New York, held February, 1881, the President of the society, in his Inaugural Address, called the attention of the society to the necessity for a change in the code of ethics, perhaps the need for an entirely new one.

The committee to whom the President's address was referred reported the following resolution :

" *Resolved*, That a special committee of five be appointed by the President, to be designated a ' Committee on the Code of Ethics,' whose duty it shall be to consider the whole question of desirable changes in the code, and who shall present to the society, at the session of 1882, such suggestions on this subject as their observations and investigations may direct."
This resolution was adopted by the society, and a committee of five was appointed. Of the *personnel* of this committee the following may be stated : Three of its members were chosen from among the older members of the society and the profession, and two from among those who had been in practice between fifteen and twenty years. Three of the members had been presidents of the society, and the other two had served on important standing committees. Three were from the northern, western, and middle portions of the State, and two from the city. Three were general practitioners, and two were specialists. From this it would seem that the various interests involved had been carefully provided for. The committee gave the subject with which they were charged careful and laborious attention during the year that was allotted to them. The views of the different members were in part elicited and circulated by correspondence, and the

views and feelings of many of the more prominent, and also of the more obscure, members of the profession were sought. After a pretty complete knowledge had been obtained of what appeared to be the prevailing sentiment of the profession throughout the State, the committee, setting private business aside, devoted two entire days to the matters under consideration. The first conclusion arrived at was, that, if the profession of the State desired a code, one should be reported that should be clear and distinct in its meaning, and one that could be enforced when necessary. The second conclusion was that the code should contain nothing that was already provided for by the laws of the State, or by such moral laws as all, whether Christian, Jew, or infidel, considered binding. This narrowed the matter to the formulation of such rules as seemed to the committee most likely to be in harmony with the sentiments of the thoughtful members of the profession, and to conduce to the best interests both of it and of the public.

The two most important sections of the code were, first, those relating to the matters which the Committee on Ethics of the New York County Society had brought to the notice of the profession, and, second, those which related to the question of consultations. In dealing with this matter, the committee carefully examined the American code, and found, as had been pointed out by the New York County committee, that it did not fully cover the ground. They, therefore, added several supplementary clauses, which made the completed article read as follows:

"It is derogatory to the dignity and interests of the profession for physicians to resort to public advertisements, private cards, or handbills, inviting the attention of individuals affected with particular diseases, publicly offering advice and medicine to the poor without charge, or promising radical cures; or to publish cases or operations in the daily prints, or to suffer such publications to be made; or, through the medium of reporters or interviewers, or otherwise, to permit their opinions on medical or surgical questions to appear in the newspapers; to invite laymen to operations; to boast of cures and remedies, or to perform other similar acts.

"It is equally derogatory to professional character, and opposed to the interests of the profession, for a physician to hold a patent for any surgical instrument or medicine, or to prescribe a secret nostrum, whether the invention or discovery, or exclusive property, of himself or of others.

"It is also reprehensible for physicians to give certificates attesting the efficacy of patented medical or surgical appliances, or of patented, copyrighted, or secret medicines, or of proprietary drugs, medicines, wines, mineral-waters, health resorts, etc."

We believe no open objection has been made to any of the provisions of the foregoing sections, except with reference to the matter of patenting surgical instruments. It is claimed by those who advocate the propriety of patenting instruments that there is really no difference between that and taking out a copyright on a book. Personally, we can not regard the matter in that light; for, if this be admitted, a parity of reasoning would indorse the propriety of patenting medicines.

In the second important matter connected with the revision of the code —namely, the consultation question—the committee felt that the gravest responsibility rested on them. In dealing with it, they believed that a correct and lasting solution would alone be reached by discarding sentiment and their own personal preferences, and considering the matter from the stand-point of actual fact. It was perfectly well known that consultations between regular physicians and homœopaths were of frequent occurrence. It was also perfectly clear that the disposition to prosecute and discipline offenders for this breach of the code had disappeared. The last case of discipline known to the committee was the Gardner case, fifteen years ago. It was also deemed probable that, since the homœopaths, by formal resolution, had repudiated their "exclusive" position, and had thus escaped the letter of the code, convictions of offenders would be exceedingly difficult; and that, if a society should convict a member and suspend or expel him, the courts would, on technical, if no other grounds, inevitably reinstate him ; and a society repeating such action would probably become amenable to the charge of contempt of court, with its attendant consequences, and possibly liable, also, in civil damages to the aggrieved party. It was almost morally certain that no prosecutions of this sort would be undertaken, except by some indiscreet person, for purposes of gratifying private malice. The heresy-hunters of a preceding generation had mostly disappeared, and there were apparently none left who felt it their duty to act as public prosecutors. It was perfectly clear to the committee that the restrictive clause of the code availed only with those who felt themselves in honor bound by its apparent spirit, while it left all others to do as they pleased, free from any anxiety as to the consequences.

Under these circumstances the committee had but two courses before them—one of which was the preparation of a consultation clause so carefully and tightly drawn that escape from conviction would be impossible, or else to recommend the abolition of all restrictions on the subject, leaving the matter to the individual consciences of all those who were interested. If the first course had been adopted, there is not a shadow of a doubt that the courts would have pronounced it "*contra bonos mores*," and void. The people of the State, as well as the legislators, had already become sufficiently indignant against the profession for assuming an attitude that appeared to them bigoted, intolerant, and inhumane. The rule of the American Medical Association was generally regarded as iron-clad, admitting of no exception ; and cases almost without number were known to the committee in which medical men had refused consultation assistance under circumstances that laid them open to the gravest charges of inhumanity, the only excuse given being that the rules of their order forbade them doing otherwise. The code of the American Medical Association, in its true and intentional meaning, is rigid and inflexible ; no matter what may be the occasion, a physician meeting or consulting with an "irregular" was liable to discipline.

The committee, therefore, deemed it both useless and unwise either to retain the rule of the American Medical Association or to recommend the adoption of a stricter one. The medico-political aspect of the question also received consideration, and from two points of view. In the first place, the effects of the exclusive attitude of the profession on the homœopathic question were duly weighed, and it was the unanimous opinion that the consultation clause of the code of the American Medical Association had, more than any other one agency, assisted the homœopaths to obtain their present position in the estimation of the public, and the abolition of this clause was the first step to be taken if it were desirable that the people should again estimate medical men according to their individual merits, rather than as upholders of this or that doctrine. In the second place, those who had interested themselves to obtain legislation on medical subjects intended to improve the status of the profession of the State were frankly informed that no relief might be expected from the Legislature so long as the profession was at war within itself. When intestine differences were healed, the State would be only too glad to do what it could to elevate and improve the material condition of medical men ; but, so long as there were factions, the State would take no action that might perhaps aid one to the detriment of the other. These reasons alone should, in the writer's judgment, have been sufficient to decide the question at issue, but there were others which appeared to the committee to be even weightier. The relation of the profession to the welfare of the community was an element that could hardly be overlooked. To exemplify this point briefly, it may be assumed that the only instance in which a homœopath would desire or ask for a consultation with a regular would be when in the treatment of a given case he had exhausted his own resources, and the patient still remained uncured. Under these circumstances, duty to his patient certainly demanded that he should seek advice and counsel from such sources as in his judgment would be best able to supply them. To this end he solicits the aid of a Thomas, a Flint, or a Sayre, believing that their larger experience in certain departments may throw a clearer light on the pathology of the case, or may enable them to suggest a more successful method of treatment than the one previously pursued. If, now, these gentlemen believe that their own methods are not superior to those commonly pursued by homœopaths, or that their skill is less than that of the physician who seeks their aid, they certainly have valid excuse for declining a consultation. If, on the other hand, the consultant has reason to believe that his experience or skill may contribute somewhat to the recovery of the patient, it would certainly seem that his duty to the individual, and, in a wider sense, to the community, was perfectly clear. To the performance of this duty there has heretofore been but one obstacle—the consultation clause of the American code.

Still another aspect of the question presented itself—namely, the right of a society to lay down any restrictive rules for the guidance of its members which interfered with the free exercise of their talents and abilities in the pursuit of their calling. It may be conceded that there exists in every organized body a necessity for certain rules and regulations relative to its organization and continuance; and it may be contended that individual members should yield some of their personal rights, if the general body to which they belong will be benefited thereby. The present question, however, does not appear to fall within either category, as it certainly will not be claimed that, if A (regular) consults with B (homœopath), C (regular) is injured thereby, or that the fellow-members of A and C receive any detriment as a body from the action of A. We, therefore, fail to see any good reason why A should be restricted in the matter of consultation when either his sense of duty or his inclinations or interests are at stake; and a rule that does so restrict him belongs to the class of rules which American citizens have always regarded as opposed to that liberty of action which is referred to in the Declaration of Independence, and guaranteed by the Constitution of the United States. It is a rule utterly opposed to the principles that underlie the National and State governments of this country.

The foregoing may be regarded as among the moral aspects of the question, and those which most certainly should be the first to be considered; and the committee were of the unanimous opinion that ordinary morality and the welfare of the community demanded that the old rule should be abolished, and the matter of consultation left to the good sense and conscience of each qualified practitioner.

The question of expediency next demanded attention, and more especially in its relation to the sectarian bodies. Would the proposed action aid them to maintain their antagonistic attitude? Or, on the contrary, would it not, by removing the chief excuse for their existence, tend to their gradual extinction?* The committee were unanimously of opinion that such would be the effect of the contemplated action.† It must not be supposed, however, that the committee overlooked the relations which the State society bore to the American Medical Association. On the contrary, these were considered most carefully and exhaustively; but since much misapprehension exists as to the relationship between this body and the various societies which are represented in it, we briefly state the facts:

The American Medical Association is a voluntary and self-constituted body, without charter or any form of incorporation, amenable to no other authority than its own will, and without power to exercise authority over any other body. At the time of its organization it adopted certain rules by which its future membership should be regulated, which rules have,

---

* The doctrinal changes that had occurred in both the homœopathic and regular schools during the past fifteen years were so great, and in such converging lines, that there was no longer any sufficient *raison d'être* for the continuance of societies whose coherence depended on dogmatic or doctrinal peculiarities, their only real bond of union being one of political defense against the aggressive attitude of the regular profession.

† The very decided confirmation of this opinion by recent events will be shown later.

from time to time, been amended and changed. The association indicates the kind of societies from which it will receive delegates,* and the terms on which said delegates will be admitted.† The acceptance of these terms by the various State and other societies simply permitted them to be represented, and to take part in the proceedings of the association.

The association is also composed in part of what are known as "permanent members," namely, persons who once or oftener have served as delegates, and who, in virtue of this fact and an annual payment of five dollars, become entitled to assume the designation.‡ The permanent members, however, do not enjoy equal powers and privileges with the delegates. They are entitled to seats, and, in a qualified sense, are allowed a voice in the proceedings. They can not, however, give practical force to any views that they may hold on topics under discussion, inasmuch as they are not permitted to vote. It will, therefore, be seen that the practical management of affairs is taken out of the hands of the older and more experienced members and left to the judgment of those who are younger, many, if not most, of whom visit the association for the first time ready to become the veriest clay in the hands of some wily and adroit manipulator.#

It will be seen from the foregoing that the American Medical Association is not in any sense a confederation of State and dependent societies, united by mutual pledges to each other, as is the case with the States forming the Federal Union, but simply a body composed of such societies as find it to their inclination or interest to conform, for the time being, to certain rules and regulations. If, therefore, a State society should place itself in a position that would prevent further continuance of its connection with the association, such action can not be considered as a *secession* in the same sense as a withdrawal of States from the Union would be. It is not a violation of any promise or pledge, but simply the severance of a connection which, in this instance, the committee believed was at the present time, and under present circumstances, a source of injury to the profession of the State.

---

* These are permanently organized State medical societies, and county and district societies entitled to representation in the State societies.

† The adoption of the association's code of ethics.

‡ At one time a single payment of five dollars and attendance on a single meeting of the association were the only requirements for permanent membership.

The constitution of the New York State Society is essentially different. In this body the delegates, in accordance with the statutes of the State, are elected for four years, one fourth of the total number being elected annually and one fourth retiring. At every meeting, therefore, there will be those who are serving in their fourth year, others in their third, and others in their second, while but one fourth of the entire number of delegates can by any possibility be men without experience in society matters. In addition to the delegates, the permanent members constitute an important factor in the society. They are chosen from among such delegates as have served three out of their four years of service, and the number that may be elected annually is limited by the statutes of the State. Permanent members have both voice and vote.

In the light of these facts and conclusions, the committee decided to report the following for the consideration of the State society:

"Members of the Medical Society of the State of New York, and of the medical societies in affiliation therewith, may meet in consultation legally qualified practitioners of medicine. Emergencies may occur in which all restrictions should, in the judgment of the practitioner, yield to the demands of humanity."

As the subsequent sections of the State code have not been specially criticised, we will not quote or make further allusion to them.

After the committee had agreed on their report, the question of publishing it in advance of the meeting was considered. The committee would have been glad to give it the fullest publicity, but they were without authority so to do. It would have been contrary to custom, and an act of disrespect to the society, for any of its committees to give to the public a contemplated report in advance of its presentation to the body that had ordered it. It was thought proper, however, to show it to some of the ex-officers of the society, and especially to those who, it was supposed, might have opposite views to those of the committee. The writer was responsible for the use of but a single copy in this manner. It was shown to an ex-president of the society, who, in a letter received the following day, commented on it as follows:

"If the spirit of the new code, which is proposed, and the spirit of the resolution which you read to me as unanimously adopted by the Royal College of Physicians of London, had governed the profession forty years ago, homœopathy would never have attained an elevation, in the opinion of any of the educated or cultivated portion of the community, as an antagonistic school in medical science. Both the profession and the public would have been saved much evil."

FIFTH ARTICLE.

From the New York Medical Journal for June 2, 1883.

AFTER the committee had agreed as to the report that should be made to the society, it directed that it should be printed in advance of the meeting, in order that it might be placed in the hands of those present at the earliest possible moment after its reception by the society. This was done that each and every member might have an opportunity to carefully consider its contents and prepare himself to offer objections to it if he saw fit to do so. The report was printed with the lines numbered, to facilitate reference when it came under discussion. It was presented at the morning session of February 6, 1883. It was duly received, the printed copies were distributed to the members, and the subject was made the special order for the evening session. During the day, and up to the time of its formal consideration, the code was the chief subject of conversation, and it is safe to say that it was looked on by the members from all sections of the State as a very decided improvement on the American code, and one that, in its practical operations, would accomplish much good for the profession in the State. The members present were apparently almost unanimous in its favor, the leading comment being that the change was one that should have been made years before.

At the evening session, the report of the committee having been read, Dr. Agnew moved that it be discussed *seriatim*. Dr. Roosa moved, as an amendment, that it be discussed as a whole, which was ordered. Dr. Roosa, taking the floor, then said : *

If the society will bear with me in a few remarks, I will at their close offer a substitute for this report. I recognize the character of the gentlemen who have made this report. Without exception, they are the honored servants of this society; without exception, they have had peculiar opportunities to learn the will of the profession in this State with regard to the code of ethics. I also recognize, Mr. President, the very great difficulty under which this distinguished committee has labored, for I remember that behind them are the traditions of a profession that believed it was necessary to bind each other with very strong legal bonds in order to prevent harm. I remember that they saw the traditions which

---

* The extracts from the discussions are taken from the official record as published in the Transactions of the Society for 1882.

were thought to be as obligatory as those of the Mosaic code, and necessary in order to promote righteousness among medical men. I therefore see their difficulties, and I honor the result of their labors. I believe that it presents a great advance over anything which has been offered to our profession up to this time. But my objection to this report is that it contains nearly every one of those things which in the progress of time have become distasteful to the profession of our day. I believe that it contains in it the very intrinsic objections which we constantly make against the code which I hold in my hands—this sentimental code of our forefathers, which tells us how our patients should behave toward us, and which enters into such innumerable details as to the relations we sustain to our fellow-men that it is impossible to believe that the authors of it thought the medical profession was entitled to any discretion in the management of its own professional affairs. I think that, if the committee had fully studied the sentiment of the profession of the Empire State, they would have wiped out the code of ethics from its beginning to its end. I believe they would have left such matters to be settled by the individual discretion and wisdom and the good faith of each man in affiliation with this society.

After further remarks, Dr. Roosa offered the following substitute for the report of the committee :

The Medical Society of the State of New York, in view of the apparent sentiment of the profession connected with it, hereby adopt the following declaration, to take the place of the formal code of ethics, which has up to this time been the standard of the profession in this State.

With no idea of lowering, in any manner, the standard of right and honor in the relations of physicians to the public and to each other, but, on the contrary, in the belief that a larger amount of discretion and liberty in individual action, and the abolition of detailed and specific rules, will elevate the ethics of the profession, the medical profession of the State of New York, as here represented, hereby resolve and declare that the only ethical offenses for which they claim and promise to exercise the right of discipline are those comprehended under the commission of acts unworthy a physician and a gentleman.

Resolved, also, That we enjoin the county societies, and other organizations in affiliation with us, that they strictly enforce the requirements of this code.

Dr. H. G. Piffard, of New York, moved that the substitute be referred to the Standing Committee on Ethics, to be reported upon next year. The motion was lost.

Dr. Squire, of Elmira, then spoke to the subject, and was followed by Dr. Van der Poel, who said :

Let us for a moment see where we stand. The special committee upon the code of ethics, after consultation, have presented this report for adoption. That report is objected to upon the one hand by Dr. Roosa, because, liberal as we have made it, it does not go as far as he thinks it should, and we have placed restrictions upon ourselves which Dr. Roosa wishes to have swept away. On the other hand, Dr. Squire thinks that we have swept away too much of our restrictions in some directions, and that we have not swept enough away in others. Now, let me state what was the spirit which governed us in making up this code as presented in this report. We reached the conclusion that it would be impossible to affect the relations of man to man, or the gentlemanly conduct and behavior between man and man. We can not make a man a gentleman unless he is made so by nature; it is utterly impossible to bind men in these relations by any code of medical ethics. For that reason we left these things out of our report. Every one of Dr. Squire's references relate to conduct between man and man, and, as we believe that no written restrictions can affect the moral character of the man, we simply say that, in our opinion, we should govern our conduct in consultations as we have indicated in the report. Dr. Roosa goes further than we do, and he wishes to restrict the medical profession only by those influences which are comprehended under the bonds " worthy of a physician and gentleman "; and says that it is utterly useless to make any obligation, and that it should be left to the moral decision of each practitioner, and if you choose to consult with any man, you are perfectly at liberty to do so. I think the time has come when consultation should be made vastly more liberal than it has been. I have grown up with all the prejudices and tendencies of a man educated at the time I was, and during my entire life-time I have not consulted directly or indirectly with a homœopath, and therefore I can speak without fear or favor. But, for a few years past, I have been somewhat removed from active practice, and I have looked over the question a little more dispassionately; and it has struck me, as well as others, that our position in this respect was painfully narrow and restricted. If we can break down those barriers and show up the homœopaths to the public, and break down the barrier which enables them to get sympathy from the public, and leave it openly with the conscience of every gentleman to go or not to go as he sees fit, I think it would be a very great advance. There are many physicians from the country here at this time, and doubtless they will support me in the statement that the instances are not few in which there are two physicians in a village—one a homœopath, and the other a regular physician, using the ordinary expression. The homœopathic physician, perhaps, has a severe case of sickness, and it becomes necessary to have consultation and advice, and the circumstances are such that he can not send away to get such consultation and advice. Now, it seems to me to be cruel and heathenish, although I have done it myself over and over again, to hang upon a miserable code of ethics and say I can not go. Such cases, I believe, should be left to the decision of the gentlemanly feeling and instinct of the man.

The present writer then took the floor and objected to Dr. Roosa's substitute, fearing that it would be generally interpreted as removing all restrictions on the conduct of medical men and lead to the most unbridled license, and, in fact, to an ethical status equivalent to that of the eclectics.

Dr. Frazier, of Camden, said:

Now, upon the one hand the laws of this State say that certain men are physicians, and make them so by law; and here we have simply an effort made to make our laws comply with those of the State of New York. By not doing so, we give to other physicians who do not belong to our school a very great advantage in permitting them to say, in different neighborhoods: " We are practicing according to the laws of the State of New York, but these men will not meet us in consultation according to the laws of the State under which we all practice." That has been the argument which they have used, and which makes them strong, and which makes us appear stubborn and weakens us. I have always been of opinion that we should be permitted to meet these men if the

3

permission could be placed in proper language; that is, if we could be permitted to consult with them without making it obligatory that we should do so.

Dr. H. D. Noyes, of New York, said: ·

I have been looking for something which would explain to me what it was that gave rise to this system of rules which we know under the name of the code of ethics, but I have not heard whence it came. It seems to me it must have had a beginning in controversies and animosities and peculiar conditions, which have to a large extent, or, perhaps, entirely, been abolished. The present status of the medical profession is one in which there is a high sense of personal dignity, and feelings of propriety among its members in their relations to the public and to each other. At the same time we all know that a considerable class of men—both those who are called regulars and those who are irregulars—are prone to do things improper and unworthy of gentlemen. Now, the code of ethics, as it stands, is doubtless intended to meet the latter class of men, but I think that our influence in controlling them is absolutely nugatory. I do not believe that the old code of ethics has amounted to anything in the way of restrictions upon them. Every physician can call to mind instances of flagrant violation of both the spirit and the letter of the code by men who have not been called to account. I well remember the feeling of surprise which came over me when the American Medical Association met in the city of New York in 1854, and I first read what is known as the code of medical ethics. I never was more struck with astonishment than with that document. It seemed to be saying things which were both humiliating and unnecessary, even at that time. I am sure that the sense of this meeting is for the abolition of that code as it now stands. . . .

. . . My feeling throughout my professional life has been, first, to study what my duty is to humanity, and, second, to consider what my duty is to the profession. It has been with me a strong desire, and continues now, to see wherein professional honor and propriety can be sustained, and, in deciding with reference to consultations, I must say that I have uniformly refused to consult with so-called irregulars. At the same time I have done so under the feeling that I was entitled to protest against it. I shall vote in favor of the substitute.

Dr. E. M. Moore, of Rochester, said:

I have had the same wonder which Dr. Noyes has expressed as to where such a code as theirs (American Medical Association code) could have come from, and I took pains at one time to investigate in that direction as to how we could have had such a wonderful composition. It was really copied almost verbatim from a treatise written by Dr. Samuel Percival in 1760, in accordance with the condition of society presented in England at that time, which was entirely aristocratic, governed by the law which regulates the relations existing between patron and client.

About seven or eight years ago this subject came up in the American Medical Association. It began to be doubted whether it was all right. I received at that time a letter from Dr. N. S. Davis, of Chicago, the father of the association, to whom this matter had been referred, and I gave him my views upon the question, calling attention to the fact that Dr. Percival wrote the article over a century ago, and expressed the opinion that we had entirely outgrown such swaddling clothes. At the next meeting of the American Association the committee brought in their report, and it was to the effect that the code of ethics was so excellent that it should be maintained. . . . I am willing to strike it all out and leave the regulation of our conduct to the unwritten law, for we have uniformly failed to apply the rules of discipline under any code or system of ethics by which we have been governed. . . .

Additional remarks were made by Drs. Mosher, Wight, Squibb, and Gray. It is a curious fact that not a single speaker defended the propriety of the consultation clause of the old code, or attempted to advocate the merits of this instrument as a whole. Many, however, thought it would be better to defer the decision of the question until after it had been brought to the notice of the American Medical Association. The majority thought otherwise, and the vote showed 52 in favor of immediate action to 18 opposed. Immediately after the vote, Dr. Roosa gave notice that at the meeting in 1883 he would move the adoption of his "substitute."

On the writer's return to the city, the almost universal expression of opinion was in favor of the action of the State society, the adverse criticisms being very few in number. On the day of his return he was met in the street by a well-known physician, who asked concerning the ethical situation. On being informed, he said : " I suppose, then, I can call in a homœopath if I want to." I replied that I saw no reason why the rule didn't work both ways, but inquired why he wanted a homœopath. He replied that he had a very severe case of scarlet fever under his care, and that the family were very anxious, and desired a homœopath in consultation ; and added that, in all probability, if he did not have the homœopath, the family would discharge him and employ the other. He stated that, rather than lose the family, he would have the consultation, provided he could not be disciplined therefor. This gentleman is now enrolled as one of the supporters of the old code.

The publication of the action of the State society permitted the medical press throughout the country to criticise it, and this they did in a manner that was not altogether in harmony with that spirit of Christian charity that the old code was supposed to be founded on. If the fasehood, hatred, and malice displayed by some of the advocates of this code are to be regarded as its legitimate fruits, God forbid that it should ever again be the supreme medical law in this State. The change in the code was a matter that affected the profession of this State only, and was not the business or concern of any one outside the State. The State society, with perfect deliberation, and with its eyes open, resigned from the American Medical Association,* because it believed that the consultation rules of that body were both morally and politically wrong, and that further acquiescence in them

---

* The severance of this connection was exactly similar in principle to that which occurs when a member, for reasons that seem satisfactory to him, severs his connection from a medical society by resignation. It is neither an act of secession, nor an act of rebellion, terms which have been used in this connection by those who should have known better.

would be to perpetuate the evils that had already accrued from their observance.

The medical press of the country, in dealing with the action of the State society, with few exceptions, failed to discuss the question on its merits. Few, if any, sought to ascertain the real motives and reasons which rendered the change necessary or desirable ; few even discussed the effect that the change would have on the affairs of the 'profession in this State. Few appeared to recognize the fact that it was the duty of the State society to watch over and protect the interests of the profession rather than to leave them to the tender mercies of the American Medical Association.

On the contrary, the majority of the press assumed to view the subject in a light which reflected most seriously on the intelligence or honesty of the profession in this State. One of the most prominent charges, and one which, after it had been first enunciated, passed current as truth from one journal to another, was the statement, wholly unverifiable by facts, that this movement had been initiated and engineered solely by specialists from sordid motives.* The large vote (fifty-two to eighteen) by which the original change was made indicates either that the great majority of the profession in this State are specialists, or else that a very considerable number of general practitioners were of but one mind in this matter.

It may be asked why the State society did not first apply to the American Medical Association before taking this action on its own account. To which we may reply that, in the first place, it did not desire to carry a domestic affair into the councils of that body, to inflict its own views on the profession of other States, to make any attempt to proselyte in ground that was as yet unprepared, or to meddle with the affairs of those beyond its jurisdiction. Secondly, it knew the history of the American Medical Association too well to expect for a moment that it would listen to any propositions looking toward the liberalization of the profession. Almost from the beginning of the history of that association its practical management and the dictation of its policy have been in the hands of one man— one who is commonly spoken of as the "Father of the Association," and who, so long as he retains his power, will continue to use it, as he has used it in the past, as an obstacle to the scientific and political advancement of the profession and the welfare of the people. For evidence on this point,

---

* All of the New York specialists who had anything whatever to do with the matter, or voted on the question, were members of the New York Academy of Medicine, which was under the governance of the American code. The action of the State society in no wise affected the ethical status of that body, which is competent to adopt any by-laws that it chooses, or any code of ethics that it desires. These members of the State society, therefore, who voted for the change simply gave freedom to those who were not members of the Academy without releasing themselves in any manner from the Academy's code ; nor have they, during the past eighteen months, made any effort, or organized any movement, to change the ethical status of the Academy. It will be seen, therefore, that the specialists were not governed by the sordid motives that have been ascribed to them. They, less than any one else, have been affected by the change.

we need but to cite the latest ethical ordinance adopted by that body—one which clearly showed the animus of the controlling element of the association, and exhibited its ever-readiness to meddle with the domestic affairs of the individual States. We refer to the action that was recently taken in reference to the University of Michigan.

At home, the action of the State society was discussed more temperately. The two most important medical journals of this city approved the change, while one or two others of minor influence were opposed. The profession of the State as a whole gave the matter but limited consideration, but as a rule a favorable one; the old code was pretty generally regarded as dead, and the recent action was simply the interment of its remains. A few months later Dr. E. R. Squibb circulated throughout the profession of the State, in his personal organ, "The Ephemeris," a protest and an argument against the change. This action was the first incitement to dissension and trouble in the State, and we can not but regard it as ill-judged and in bad taste. Dr. Squibb is not a practitioner of medicine, and has not for many years been placed in a position that would enable him to have a practical knowledge of the questions at issue; and, even if he had, he should have left the discussion to those whom it immediately concerns. As a recognized member of the profession, however, he had a perfect right, of course, to discuss the matter; but it would have been better, we think, if he had done so in one of the public medical journals, in which his views and opinions might have received reply. As it was, he of necessity had the entire argument to himself, and was enabled to give his personal views the widest distribution that he chose.

A journal in a neighboring city, with even less motive, endeavored to excite dissension among the members of the county societies in this State, and incited them to seditious action, urging them not to accept the edict of the State society, but to adhere to the American code, apparently unaware that such action by the county societies would be absolutely null and void.

Our Southern friends also, with great unanimity, censured the profession of the State, and even suggested that the malcontents form a new State society, forgetting the old rule in physics that two bodies can not occupy the same space at the same time. The distance which separated them from us and the different conditions which surrounded them are sufficient reasons for a misapprehension on their part of the propriety and necessity of our action. We think, however, that they might have been a little more charitable in the expression of their views. They frankly charged us with desiring to affiliate with the lowest quacks and charlatans, just as in the old antebellum days they charged every abolitionist and opponent of slavery with the desire to marry his daughter to a "nigger." Long accustomed as they were to the idea that slavery of the body was right and proper, it may take them some years before they understand why it is that we in the North are unwilling to longer accept the slavery of the mind of which the old code represented the bonds.

Shortly after the meeting of the State society in 1882, an attempt was made to bring up the matter of the code in the New York Academy of Medicine. Notice was given that the action of the delegates from the Academy who had voted for the new code would be submitted to the Academy for its approval or the reverse. This was done by the friends of the old code who hoped to secure an expression of opinion adverse to the action of its delegates. At the meeting of the society at which the question was to be considered, Dr. Austin Flint, Sr., first claimed the floor, and, in a few brief but eloquent words, urged that the question be not brought up in that body, and moved that all action on it be indefinitely postponed, on the ground that the scientific interests of the association were too important to be hazarded by the introduction of medico-political questions of this character. This action was regarded by the opponents of the old code as almost a pledge that the question would not be raised again. They preferred to abide by rules and by-laws which were distasteful to them, and to gauge their conduct in accordance with ordinances regarded as oppressive rather than to risk the future harmony and welfare of the Academy. During the succeeding months the advocates of the State code rested on their oars, but the supporters of the American code were busily engaged in manufacturing opinion throughout the country adverse to the action of the State society. The most active agencies in this direction were the "Medical News" and the "Ephemeris," both of which endeavored to incite sedition in the county societies, urging them to repudiate the action of the State society, and to instruct their delegates to vote for the repeal of the State code.

Thirty or more of the county societies followed this advice, and instructed their delegates to vote for the repeal of the State code. The societies that took this action were only those entitled to a limited number of delegates, and the aggregate vote thus obtained hardly counterbalanced the votes of two or three of the more populous counties. In none of the larger counties was this action taken. It is true that Kings County, entitled to twelve delegates, did at one meeting instruct its delegates, under the urging of Dr. Squibb, to vote for the repeal of the State code, but this action was rescinded at a subsequent meeting of the society.

In New York County, it being the year for the election of delegates, upward of forty nominations were made at the September meeting. Of these candidates I do not recollect the names of more than two or three who belonged to the old code party. From the forty nominees, twenty-four (the number prescribed by law) were to be selected. At the election in October, the entire number elected were men opposed to the old code. This was not in consequence of any special electioneering, as from the code standpoint it mattered little which of the forty (with two or three exceptions) were elected. Those elected owed their election to the fact that they were suitable persons for the position, and were sufficiently popular. One or two of the unsuccessful candidates were men equally suitable and personally popular, but, their pronounced position in favor of the old code being known, they polled a very small vote.

After this, very little thought or attention was given to the code in this city, much less, in fact, than in other parts of the country, our neighbors appearing to be really more solicitous about our welfare than we were ourselves. Ethical matters slumbered, and it was not until the stated meeting of the county society in January, 1883, that the question was again opened. At this meeting, Dr. L. A. Sayre moved that the delegates of the society be instructed to vote for the repeal of the State code. After discussion, this motion was lost. Not satisfied with this expression of opinion on the part of the society, the old code advocates procured the calling of a special meeting for the announced purpose of obtaining a vote on the question. At this meeting the question of instructing the delegates was again raised, and a motion to that effect was lost by a vote of 60 yeas to 147 nays.

SIXTH ARTICLE.

From the New York Medical Journal for October 6, 1883.

THE seventy-seventh annual meeting of the Medical Society of the State of New York convened February 6, 1883, and was opened with the inaugural address of the President, Dr. Harvey Jewett, of Ontario County. From this address I extract the following:

At the annual meeting in February, 1881, this society appointed a committee of five, from among the most distinguished medical gentlemen of the State, to consider and revise the old code of medical ethics which had governed our action for nearly forty years. In conformity with the instructions given this committee, they presented their report at the annual meeting in 1882. At the same time a substitute was offered to this effect: that we abolish all restrictions relative to the practice of medicine, as superfluous and unnecessary in the presence of the unwritten or higher law, leaving all ethical questions to be settled by the gentlemanly instincts of the profession. The report of the committee, as well as the substitute, was printed and placed in the hands of all members of the society who desired a copy, that they might examine and vote deliberately and understandingly upon the changes reported for their consideration and adoption. After a general discussion, in which all present had an opportunity to express their views, the report of the committee was adopted by a large majority. The new code has not been received by the profession or the medical press, in this and in other States, with cordiality or favor, but, on the contrary, by the most outspoken and emphatic opposition. The county societies, at their first meetings, expressed their surprise at and disapproval of the new code adopted by a majority of their representatives, as unbecoming the dignity of the profession, and as revolutionary in its nature and "disorganizing in its tendency." A year's consideration, a calm and dispassionate discussion of the matter, have greatly modified the views of the profession in reference to the objectionable measure, and I trust a more conservative sentiment exists to-day than at the time of its adoption.

The American Medical Association, at its annual meeting at St. Paul, in June, 1882, refused admission to the delegates from the Medical Society of the State of New York, because they failed to recognize some of the provisions of the old code which had controlled their action for so many years, and had taken the liberty to substitute what was deemed a more progressive and liberal spirit in reference to established rational medicine as it exists at the present time. The objectionable clause in the new code consists in the   sion of consultation with any legally qualified practitioner of medicine as not derogatory to the interest and dignity of the profession, in cases of emergency, or where such aid is required upon the broad ground of common humanity.

The advocates of the new code assert that this is merely permissive ; that no one is under obligation, expressed or implied, to meet an irregular practitioner in consultation, unless he prefers to do so ; but in certain cases it would be illiberal, inhumane, and contrary to the spirit of the age, to withhold professional aid because of "difference of opinion in creed or belief." The attention of the society at this meeting is directed to a consideration of the merits of this subject, to confirm, modify, or abolish the new code, as in their wisdom and judgmen ᵜ they may deem most conducive to the welfare, dignity, and interests of the medical profession of the State of New York.

At the conclusion of the President's address the standing committees of the society were announced, and after that communications from county societies, as the first order of business, were called for. The following were presented :

MEDICAL SOCIETY OF THE COUNTY OF WESTCHESTER,
KATONAH, *February 3, 1883.*

To the State Medical Society :

The following action was taken by the society at its annual meeting in 1882 :

Resolved, That this society reaffirm its loyalty to its parent body, the American Medical Association, and thus declare its adherence to the code of ethics prescribed by that body as a guide in practice. It strongly deprecates the action of the State society, and maintains that such action is as unworthy as it is revolutionary ; and that the adoption of such a code under such circumstances could result only in confusion and dishonor.

Resolved, That these resolutions be submitted to the State Medical Society at its next annual meeting through our regular delegates.

Carried. Affirmative, 28 ; negative, 2.

(J. G. WOOD, *Secretary pro tem.*)

J. FRANCIS CHAPMAN, *Secretary.*

ROCHESTER, N. Y.

At the annual meeting of the Monroe County Medical Society, held in the Common Council chambers at Rochester, N. Y., May 31, 1883, it was

Resolved, That it is the sense of the Monroe County Medical Society that the code of ethics be repealed, and that the secretary notify the Medical Society of the State of New York of the action taken.

WILLIAM F. SHEEHAN,
Secretary of the Monroe County Medical Society.

OSWEGO, N. Y., *January 31, 1883.*

To the New York State Medical Society :

At the annual meeting of the Oswego County Medical Society, held in Oswego, June 13, 1882, the following resolution was adopted, and the delegates to the next meeting of the New York State Society were instructed to bring the same before the society :

Resolved, That, in regard to ethics of consultations, the true rule of our profession is that, while we should be free to visit the sick under all circumstances and under whosever care, it is unworthy of us to call in consultation any but regular practicioners.

P. M. DOWD, *Secretary.*

After the transaction of some further business, Dr. E. R. Squibb, of Kings County, offered the following:

Whereas, The Special Committee on the Code of Ethics, in its report at the last annual meeting, recommended a change in one part of the code which was more in the nature of a revolution than of a revision, and, therefore, may be more radical than was expected or desired by the constituency of this society; and,

Whereas, That report was adopted at a session wherein only fifty-two members voted in the affirmative, and thus legislated for the entire profession of the State on a subject of vital importance in a direction which may not have been anticipated or desired by the profession at large; therefore,

Be it resolved, That all the action taken at the annual meeting of 1881, in regard to changing the code of ethics, be repealed, leaving the code to stand as it was before such action was taken.

Resolved, That a new Special Committee of five be nominated by the Nominating Committee of the society, and be appointed by the society to review the code of ethics, and to report at the annual meeting of 1884 any changes in the code that may be deemed advisable.

Resolved, That the report of the committee be discussed at the meeting of 1884, and be then laid over for final action at the meeting of 1885.

These resolutions were made the special order for the evening session. At this session the society went into committee of the whole, and Dr. Squibb opened the discussion, maintaining that the action of the last annual meeting upon the subject of the code of ethics was contrary to the plan of the organization of the society, and to the letter and the spirit of self-government by majority rule, and, therefore, ought to be reversed.

Dr. Squibb then went on to claim that the code of ethics was a part of the constitution of the society, and analogous to the constitution of the State, and that any amendments to be made in it should first be approved by the county societies.

Dr. Roosa, then taking the floor, said:

The Medical Society of the State of New York, as one of its inherent rights, has the power to make its own by-laws, and, by statute law, it has the right to call to the bar any county society which may refuse to cause *its* by-laws to conform to those of the State Medical Society. The argument, under the circumstances under which we meet to-night, is utterly absurd. What has happened? It is true that instructed delegates have come here from several county societies, and it is also true that from these very counties have come men who have been made permanent members, and who are entirely at variance with the instructions given to the delegates. It is a most unwarrantable doctrine that the city of New York, with its nearly two millions of people, and nearly two thousand regular physicians, that the county of New York, representing a constituency something like that of twenty counties in this State, is not to have her full and proportionate voice in the discussion and decision of any question which comes before us here. Because she has the misfortune to be a city, and a large city, is her vote to be counted only equal to that of Alleghany? I believe that this society is of one accord that the argument of Dr. Squibb is not sufficient for it, and

that the State society is prepared to-night, whatever it may have been on previous occasions, to settle this question for itself, without referring it back to the
counties. The State society undertook this action of revising the code, not as
has been charged very frequently during the year, not at the suggestion of specialists, not by any arrangement beforehand. If there ever was a spontaneous
convention on the face of the earth, if there ever was a convention which represented its constituents, it was the annual meeting of the Medical Society of the
State of New York in 1882, and it has been equaled in the annals of this society
only by the immense meeting of to-night. At that meeting this society not only
passed the revised code, but, without caucusing or consultation with any person
as to whether they could or would support it, it passed a resolution much more
radical, which had not been presented to any person, except one, before it was
offered to this society at that meeting. That meeting of the society did represent its constituency perfectly well, and so does the meeting of the society to-
night. There was no unfair action of any kind in the meeting of last year, whatever may have been the statements from any source. It was an open discussion,
and the distinguished gentleman who has opened the debate had his full say, and
he was unable to convince even one third of the meeting that his views were
correct. It is assumed in the argument of the gentleman that we have such a
union with the American Medical Association that we are compelled to ask that
association before we make any change in our by-laws. Perhaps that question
has been sufficiently answered ; and a large number are present to-night who did
not hear the discussion which took place this morning on that part of the subject. Let us understand ourselves distinctly. We recognize no allegiance to
the American Medical Association except that of fraternal relations, and in case
they refuse to admit our delegates, as they refused to do last year, this relation
is dissolved. That association is not an incorporated association. If we ever
subscribed to its code, we repealed that subscription last year. The American
Medical Association has not taken a position in the medical world to be com-
pared with this society of which we have the honor to be members. There is
no secession in this business. There is no States rights in our action. If the
union of these States was no more than that which exists between the American
Medical Association and this State Medical Society, there never would have been
a rebellion. There would have been no need of one. Each State would have
been independent, as we are now of the so-called National Association. The
gentleman lays great stress on the adjective "revolutionary." Revolutionary!
We are not afraid of that word. All the advances in the world have been made
by revolutions; but revolutions are never revolutions except as they are mouth-
pieces of the people, and a revolution in this society will never be successful
unless we represent the voice of the medical profession of the State of New
York. Not of us in New York can any charge of misrepresentation be made.
The county of New York comes up here with its hands untied and without fet-
ters, and any member is at liberty to vote as his conscience may dictate. The
county of Kings, thanks to great effort in opposition to the gentleman who has
just spoken, is also unfettered upon this floor. But, I am sorry to say, I am
addressing some gentlemen who never had an opportunity to listen to arguments
on the side of those of us who believe that an advance of the profession will be
most effectually promoted by our assistance to the resolutions offered to-night,
and they come up here bound and directed as to what they shall do. I consider
these instructions as utterly illegal ; and, when I had the honor to be president

of this society, I ruled that these instructions were invalid, and one gentleman from New York violated his instructions, and he was never disciplined, although he was threatened. No, Mr. President, not upon us can any charge of revolution be fastened except that which is similar to what has emancipated many a country, and which will emancipate the State of New York. The few other arguments which have been advanced against the new code are easily answered. It has been assumed by the friends of the old code that we have played completely into the hands of the homœopaths and the eclectics, if consultation with any class of legally qualified practitioners be allowed. Now, if you are not willing to trust the ex-presidents of this society, who, with very few exceptions, are entirely in favor of this expression of freedom in consultation, and the committee, not of specialists, but of a majority of general practitioners who drew up this code—if you are not ready to trust them as to whether we are going to surrender in any such way, then I have misunderstood this society.

It has also been said that this is a medical question, and that it can not be in any manner understood by men outside of the medical profession. If this was a question as to the value of iodide of potassium, or sulphide of calcium, or the sulphate of quinine as agents for controlling the symptoms of disease, then none except men like ourselves, who have received a medical education and have had experience in its practical application, are competent to decide it. But it is not a question of drugs or drugging. It is a question of ethics, a question of man's rights, his relations to his brother man, and his entire conduct toward the people of this community. The entire sentiment and conduct of the people is against this restricted trades-union clause in the American Medical Association code, and they have a right to their opinion, and are competent to give an opinion upon this question. The old code of our profession has made us the laughing-stock of educated men. We claim for ourselves, not the privilege of affiliating with quacks, but of giving our advice wherever it is asked for.

If we act simply as benefactors to our own kind, no matter if we stand alone for the next hundred years, we shall be right, and the Medical Society of the State of New York can afford to smile at those who refuse it fellowship.

Dr. II. G. Piffard, of New York, then said :

I desire to throw a little light upon one point raised by the gentleman from Kings. He drew an analogy between the constitution and by-laws of this society and the constitution and laws of the State. His special effort is to show that the code of ethics was virtually the constitution of this State society, and that it could not be altered except by the consent of the constituencies from which the society is recruited. This view, I think, is erroneous. The code of ethics we have adopted is, and always has been before, regarded as a by-law simply. The gentleman from Kings County seems to think that we derive our authority from county medical societies, that we have no authority over them, and that our by-laws are subject to their revision, instead of their by-laws being subject to the revision of the State Medical Society. In that the gentleman is absolutely mistaken. He quotes from a certain law enacted in 1813, which gives us the power to make certain by-laws, and also gives county medical societies power to make certain by-laws, but he overlooks the fact that in 1866 another law was passed which enabled the State society to control the by-laws of county societies. . . . In other words, county societies are amenable to this society, not this society amenable to the county societies. . . .

Dr. H. R. Hopkins, of Erie, then addressed the Committee of the Whole.*

Dr. Didama, of Onondaga County, next spoke. Referring to the American code, he said :

. . . This code is the one which we adopted on condition of representation in the American Medical Association. If we repeal it, then we have no rightful representation in that association. It was repealed by a few, but their action was not the expression of the great mass of the profession of this State, only fifty-three persons voting one way, and they did not represent the opinion of four or five thousand regular practitioners of medicine in this State. There is a little complaint that our delegation—which was sent after our secession, if you choose to call it so, our cutting ourselves loose from the American Medical Association—was not received with respect and open honor. But I think every fair-minded man must allow that the association could have done nothing else. They were bound to reject the delegates sent from a society which had repudiated the code of ethics established for the government of the entire profession. The question is, Are we, the medical profession of the State of New York, prepared to cut ourselves loose from the American Medical Association?

Dr. Didama further stated that he considered a consultation with a homœopath, with a person who believed in the efficacy of the so-called dilutions, as conniving at a fraud.† In closing, he said :

With this I shall end, saying that a consultation with certain persons is derogatory to the medical profession, and that it is derogatory because those who do it are simply perpetrating a fraud.

Dr. Rochester, of Erie County, then spoke :

I arise with a full consciousness and appreciation of the soberness of this discussion to-night, and I hope that anything which I may say will be entirely free from personality. We have to look to common sense in this matter. I have been looking over this paper (State Code), and I have not seen any line of it which tells us what is to be gained by this proposed modification, except that broad humanity requires us to meet everybody who calls upon us. This I would say is simply a reflection on the medical profession throughout the length and breadth of the land, for there is no emergency, no casualty, no case of distress or anxiety to which medical men do not always go under any circumstances, without expectation of reward or remuneration. . . . Now, sir, I am a permanent member of the American Medical Association, and I have been for a long time a permanent member of that body, as I am of this society, and I am proud of it, and I should be sorry to give it up. But I will say if this new code passes I will give up my membership in the State Medical Society sooner than my permanent membership in the American Medical Association. . . .

Referring to mixed consultations, Dr. Rochester said :

We meet, we talk, but do not agree in therapeutics, very likely not in diag-

* As Dr. Hopkins's remarks have already been published in full in the columns of the " N. Y. Med. Journal," they are here omitted.

† See " N. Y. Med. Journal " for August 18, 1883, pp. 177, 178.

nosis, and the people are satisfied; but how is it with the patient? Does the patient get any benefit? Not at all. We say we can not see anything to do different from what is being done, and if we did suggest anything it would not be carried out. It is impossible for any such thing to take place. We can not do it without degrading ourselves. We maintain, then, our first position—that we are kind, generous, and liberal to all those who call upon us, and always have been, and I do not see any possible advantage which can come from this modification. Perhaps we are mistaken. Now, we know that while they are carrying these colors they are giving the very drugs that regular practitioners do, except that sometimes they give a little more.

Dr. Seymour, of Rensselaer County, then spoke to the question, but his remarks were too voluminous to be given in full, and will hardly bear condensation. We must, therefore, refer the reader to the official report. Referring, however, to certain members of the profession residing in a neighboring county, he said: "This thing will not do, and if you come up here to strengthen the hands of these men against us, we will arraign you before this society, and kill you off professionally; and, if you are backed up by your *confrères*, we will twist their necks off too. That cock will not crow." During Dr. Seymour's remarks he was interrupted by the receipt of a telegram from Dr. L. A. Sayre of New York. Relative to this telegram Dr. Seymour said: " It is, perhaps, under the circumstances, hardly in order, but Dr. Sayre's name was mentioned, and then it was stated that he had met with homœopaths, and confirmed the new code, and violated the old code, and I took the liberty of sending a telegram to Dr. Sayre, telling him that the charge had been made; and I understood that a letter proving the charge would be read at this meeting, and called upon him to vindicate himself, and I got this telegram from some one in his house: 'Dr. Sayre has been confined to his bed for two weeks, and it is impossible for him to be moved at present. He says: "I saw in consultation Dr. Baldwin, who was treating the patient most heroically all through. He had not diagnosed the case, and afterward I learned that he was a homœopath, although from what I saw of him no one would suspect that he was. He was at the time using hypodermic injections of morphine, and in no respect carrying out the principles of Hahnemann." ' "

Dr. Seymour further said: " Once or twice in my life I have violated the code myself, and been to consult with homœopaths—once in a case of placenta prævia, where the woman was bleeding to death. After I had righted things up, and got the woman so I thought she would live, I turned around to this homœopathic gentleman and said to him, 'You have abundant time now to consult with one of your own kind, and I will not trouble you any further.' For this consultation I was condemned, and I had to vindicate myself upon the principle of humanity."

This concluded the arguments on the question before the society. After some parliamentary skirmishing, the question was taken on Dr. Squibb's resolutions. The result of the vote showed 99 ayes to 105 nays, and the resolutions were declared lost. To carry them would have required a two-thirds affirmative vote, which in the ballot cast would have been 136.

The result of this ballot exhibited one fact with great distinctness—namely, that the majority of the representatives of the profession of this State were not in favor of a restoration of the old code. It had been repeatedly predicted by the hostile press outside the State that the action of the State society of the previous year would be reversed at this meeting. The vote showed, only too clearly, that, despite the exertions that had been made in behalf of the old code, and despite the abuse that had been heaped on us from without, the profession of the State were thoroughly convinced of the evil effects of the code in the past, and were not going to submit to them in the future, even at the expense of loss of representation in the American Medical Association.

Subsequent to the announcement of the vote, Dr. Roosa moved the adoption of his " substitute," which, on motion, was laid over until 1884. Dr. J. G. Adams entered a protest, as delegate from the New York Academy of Medicine, against the action of the State society. This protest was clearly an impropriety if offered in behalf of the Academy, and should not have been received as an expression of the Academy's feelings and views, inasmuch as a majority of the delegates of the Academy who were present voted with the majority. It may further be stated that the Academy had not expressed its views on the subject, and, as a curiosity in the matter of society by-laws, it may be stated that the members of the Academy have no direct voice in the selection of the delegates that are supposed to represent them, either in the State society or in the American Medical Association.

On the second day of the session the Committee on Legislation made its report, and during its discussion the question of ethics was incidentally revived. The committee having asked for an appropriation of five hundred dollars, for the purpose of procuring legal assistance, with the view to desirable legislation, Dr. Hopkins, of Buffalo, spoke in favor of the motion to adopt the report, to which Dr. Van de Warker, of Onondaga County, replied as follows:

I did have great hopes for the cause of medical education in this State, notwithstanding the fact that nearly every attempt this State has made to regulate the practice of medicine has been a terrible failure—so terrible that, if medical men attempt to make laws, it is to be hoped that they will be such laws as will be of benefit to the profession. The law of 1880, which legalized every quack, was a deplorable failure, and the law of 1874, which gave the medical societies of this State certain powers, was another terrible failure, and, every time this society has attempted to dabble in medical law at all, the profession at large has deplored the fact.

Dr. Sturgis: I would ask the gentleman from Onondaga if any effort has been made in his county to prosecute illegal practitioners?

Dr. Van de Warker: An Indian doctor rode through our principal streets the other day, adorned with war paint and feathers, and he registered in the

Clerk's office, and is now considered a legalized practitioner, and he stands on the same footing with the other members of the profession.

Dr. Sturgis: Has any attempt been made to prosecute illegal practitioners in your county?

Dr. Van de Warker: Prosecution was not attempted in my county. . . . The legislation of 1880 was not to protect the regular practitioner, but to protect quacks, just as the code indorsed last night was not for the regular profession, but for quacks.

Dr. Sturgis: I do not think the gentleman can speak fairly of a point which has not been tried in his own county, and in which he has not had any experience. At all events, what he states is at variance with what has been the experience of the Medical Society of the County of New York, and the same is good for every county medical society in the State. When that law passed, the physicians of New York made up their minds to make a fair trial of it, and to determine wherein it was deficient, and to try to remedy its defects. We went to our county medical society and told them that the question was simply one of money, and that we needed money to employ legal counsel. We said to them, If you will give us your support, we will carry out the provisions of the law. The result has been that sixty suits have been brought, and in only three has the society failed to establish its case, and in every offense the man has been fined and driven out of practice. We got hold not only of irregular practitioners, but we have our hands on the throat of one of the colleges of this city. When the gentleman says that the law of 1880 protects quacks and protects irregular practitioners, he makes a mistake.

Dr. Smith (Secretary of the Society): There is an opinion, widely prevalent in the medical profession of this State, that the mere fact of registration in the county clerk's office makes a man a legally qualified practitioner. It does no such thing. The law requires the legally qualified practitioner in his registration to state the authority under which he claims to be qualified, and a person who has no legal right to practice medicine, if he registers under the law of 1880, will often furnish evidence in the statement he makes in his registration that he is not legally qualified; so that that law, instead of protecting quacks, often causes them to furnish proof whereby they can be convicted of practicing illegally.

The introduction of the questions relating to the law of 1880 into the code controversy, by the supporters of the old code, was irrelevant and uncalled for, as it had no bearing on the real issues under discussion. One may, perhaps, excuse words spoken hastily in the heat of debate, but we can not so readily overlook misstatements made in the calm seclusion of the sanctum. The following references to the law of 1880 are taken from the "Ephemeris" for May, 1883, pp. 279 and 280:

"The law entitled An act to regulate the licensing of physicians and surgeons, passed May 20, 1880, through the efforts of the *New York County* Medical Society," etc. (italics our own).

The fact is that the New York County Society had nothing whatever to do with this law. It was passed through by the efforts of a committee of the State society appointed for the purpose, and the entire expense of its passage, amounting to a little less than fifty dollars, was borne by the State society. The committee itself was composed of one member from Albany, one from Kings, and one from New York counties.

"This authorizing and licensing registry law, which, seen now in the light of more recent action, appears as the first public step taken in this no-code movement, levels all inequalities, and ranks the best names in the profession with those qualified for no profession and undeserving of recognition, whose lack of qualifications must be all the more dangerous to the public welfare for being legally authorized and licensed. This class, though legally authorized in a roundabout way, through diplomas and certificates of bodies incorporated under a general act, would never have been legally recognized and licensed but for this registry law, and the harm done by thus recognizing a large number will far overbalance the good of preventing the registry of a few, or the prosecution of a few who may be so incautious as to register fraudulently."

With reference to the foregoing, we are compelled to say that we do not remember to have ever read two consecutive sentences in which were to be found so many errors as to fact, and language so well calculated to lead to false inferences. If the writer in the "Ephemeris" had taken a little care to ascertain the facts, he could hardly have had the hardihood to refer to the registry law as the "first step taken in this no-code movement." The no-code movement, as we understand it, originated in a resolution introduced by Dr. Roosa, at the meeting of the State society in 1882. This movement has gained a considerable following, but, so far as we are aware, not a single one of the supporters of this movement had any hand or part in the passage of the Act of 1880. We further say that of those who did give their time and exertions to the furtherance of this registry law, not a single one has since appeared as an advocate of the no-code movement.

"Levels all inequalities." Every citizen, before voting at a general election, must in this State register his qualifications; but we fail to see that this brings down the statesman to the level of the pot-house politician, or the learned and virtuous to the level of the ignorant and criminal. In one respect only, not in "all," does it level. Just so the medical law levels in but one respect only, and in a very necessary respect, as it is the only means by which the State or any one else can learn the number or the qualifications of those who are legally authorized to practice within the borders of the State. The writer of this does not feel himself specially degraded by the fact that his name is on the same list with the names of physicians whom he may deem of inferior professional quality, any more than he does that his name goes on the same polling-list with those whom he regards as politically inferior.

The sentence, "This class, though legally authorized in a roundabout way through diplomas and certificates of bodies incorporated under a general law, would never have been legally recognized and licensed but for this registry law," etc., will bear a little analysis. It implies, first, that there exists a class of practitioners who should never have been legally author-

4

ized to practice. With this sentiment we agree heartily, but the reader should be made aware that the only incorporated bodies that granted these legal authorizations under "a general law" were the State and county societies acting in accordance with powers granted them by various statutes passed between the years 1806 and 1874. The reader might also have been informed that the registry law of 1880 revokes these powers so long possessed by the county societies, and which they in so many instances grossly abused. In the portion of the sentence that we have quoted there is a curious contradiction. The writer admits that a certain "class" were legally authorized by certain "incorporated bodies," and then says that they "would never have been legally recognized and licensed but for this registry law." The fact is, the registry law did not legally authorize a single person to practice medicine who at the time of its passage was not already legally authorized in virtue of earlier laws (with an exception to be noted in a moment). The terms of the act are sufficiently explicit, and no misconception of their import should have arisen in the mind of any one who had read them. To make this perfectly clear, we quote the words of the act, italicizing the portions that bear on the present question.

A person shall not practice physic or surgery within the State unless he is twenty-one years of age, and either has been *heretofore authorized so to do pursuant to the laws in force at the time of his authorization,* or is hereafter authorized so to do as prescribed by chapter seven hundred and forty-six of the laws of eighteen hundred and seventy-two, or by subsequent sections of this act.

Every person *now lawfully* engaged in the practice of physic and surgery within the State shall . . . . register.

After the passage of the act, graduates in medicine only could commence the practice of medicine in the State. The exception that we alluded to a moment ago is in the case of medical students who had been in practice for ten years. These latter were accorded an exemption from some of the provisions of the act for a period of two years from the date of its passage. We doubt if there have been six persons in the entire State who availed themselves of this exemption.

SEVENTII ARTICLE.

From the New York Medical Journal for November 24, 1883.

A FEW days before the meeting of the State society in February, 1883, a prominent homœopathic physician of this city said to the writer that, if the society stood by the new code, he, and probably other members, would resign from the homœopathic county society, and abandon their special designation; but that, if the State society re-enacted the American code, thus showing that the old spirit of intolerance still dominated the profession, he should not leave the homœopathic society, fearing, with others, that it would be still necessary for the protection of their interests to keep up a separate organization. As soon as it was known that the old code had not been restored, and that the old-code party were in the minority, this gentleman and two other well-known homœopaths severed their connection with the homœopathic society. A month later, four others in like manner resigned and abandoned their sectarian titles. It seemed probable that this break from the homœopathic ranks would have greatly increased, and, in the writer's judgment, fully one half of the members of the homœopathic society would have abandoned sectarianism had it not been that the old-code party made renewed efforts for supremacy, and impressed many with the belief that they would ultimately succeed in restoring the old code. The bold front and the assurance of success assumed by the advocates of the American code put an immediate stop to resignations from the homœopathic organizations and delayed their disintegration.

The months of February and March of the present year were devoted by the supporters of the old code to the perfection of an organization, the purpose of which was to restore, if possible, the old code in this State. Such an organization was formed, and its efforts during the year have borne fruit, as we shall see later. During the month of March, Dr. Austin Flint commenced a series of papers in this journal on "Medical Ethics and Etiquette," which were a commentary on the American code as viewed from the standpoint of its supporters. Of this commentary we shall examine but a single portion, that relating to the subject of consultations. As we all know, the rule of the American Medical Association reads as follows: "But no one can be considered as a regular practitioner, or a fit associate in consultation, whose practice is based on an exclusive dogma," etc. Concerning this, Dr. Flint says: "The foregoing section has of late been made the subject of much discussion. Of the entire code, this section alone has occasioned dissension."

Dr. Flint is here mistaken. The new-code party, or, to speak more strictly, those who drafted the new code, were dissenters from the old for more reasons than this. When they found that prominent members of the profession, including many dignitaries of the American Medical Association, were the direct promoters of quackery and the use of secret nostrums, through the testimonials given in support of them, and when they found that the American code was apparently unable to repress these abuses, they endeavored in the New York State code to find an effective remedy. An examination of this code, especially its first section, will show how this difficulty was met, despite the fact that an effort of the same kind encountered defeat at a recent meeting of the American Medical Association. This certification of the value of nostrums by prominent members of the profession we personally consider as one of the most unfortunate developments of the last few years. It is true that some members of the profession in this city fell into the traps laid by cunning manufacturers, but the prompt action of the County Society checked the further extension of this evil. This was effected by the passage of a special resolution, as the Committee on Ethics found that the American code was defective on this point. When, in 1882, the attention of the American Medical Association was called to the abuse in question, its Judicial Council refused to make any provision for its abatement, fearing, perhaps, to cast any reflection on those of its prominent members who were, or who had been, advancing their own interests at the expense of the mass of the profession. The first section of the State code we personally regard as the most important, and we would be perfectly willing to strike out all that follows if by so doing we could secure harmony on the questions now at issue.

Dr. Flint further says: "The writer of these remarks is one of many who think that the code is here open to objection, not, however, in spirit or intent, but in phraseology." From this it would seem that Dr. Flint approves the sentiment or spirit of the consultation clause, but does not approve of the language in which it is clothed. Let us, therefore, consider these points. The intent of this clause appears to be the prohibition of consultations with certain persons in consequence of their methods of practice, founded on a belief in the value of a special exclusive dogma, together with the rejection of certain aids approved by the regular profession. Dr. Flint, however, a little farther on, says that a practice based on an exclusive dogma is not valid ground for an objection to consultation. "Any physician has a right either to originate or adopt an exclusive dogma, however irrational or absurd it may be." We must here confess our inability to reconcile the last two sentences that we have quoted. That Dr. Flint should say that he approves of the "spirit or intent" of the restrictive clause in the code, and a moment later say that the adoption of an exclusive doctrine is not valid ground for refusing to meet a practitioner in consultation, certainly appears to us discrepant and inconsistent.

The code having forbidden consultation with certain persons, let us ascertain, if possible, what persons are intended. On this point Dr. Flint says: " At the time when the code was adopted by the American Medical Association, the irregular practitioners, so-called, were for the most part uneducated men, whose practice was not only based on an exclusive dogma, but professedly to the rejection of the accumulated experience of the profession, and of the aids actually furnished by anatomy, physiology, pathology, and organic chemistry. They were steam-doctors, or Thomsonians, botanical, or herb doctors, eclectics, and the like. A system of practice based on the dogmas of Hahnemann had not then secured a hold on popular favor. A considerable number of those who became homœopathic practitioners, as they are termed, were from the ranks of the medical profession, and had received a regular medical education. Since the adoption of the code, this system has obtained a legal recognition. It has its societies, colleges, and journals. The homœopathic practitioners are an organized class, distinct from the regular profession. They are candidates for practice on the ground of a radical distinction in their therapeutical system, and it is on this ground that patients elect their services. Meanwhile, other systems in antagonism to the regular profession are comparatively insignificant as regards the number of practitioners and of patients."

Although the foregoing would seem to imply that the anathema of the code was directed as much, if not more, against the various nondescript practitioners of the time as against the homœopaths, the wording of the code itself would almost to a certainty indicate that it was specially intended to prevent consultations with the latter, as none of the other practitioners had even the pretense of an exclusive dogma. Dr. Flint, however, is, we think, in error when he states that these practitioners had not at that time acquired much hold on popular favor. As early as 1844 they had acquired sufficient hold to enable them to secure the repeal of the most important section of the Medical Acts of 1827, which, as we have already shown, opened wide the gates of the State to all forms of quackery.

Dr. Flint says: " Since the adoption of the code, this system " (homœopathy) " has obtained a legal recognition." This legal recognition we believe to have been the direct consequence of the code, and that it would never have been obtained except for the occurrences that grew out of the operations of the code. This is, of course, purely a matter of opinion, but in the present instance is based on a careful and extensive reading of the controversial literature of those days.

Continuing his commentary, Dr. Flint says: " It is fair to conclude that the framers of the code had no feeling of illiberality, and no intention to interfere with the practice of medicine, under any circumstances, in the cause of humanity. The code declares explicitly that in consultations the good of the patient is the sole object in view, and enjoins against declining consultations on the score of fastidiousness. The restrictions of the code are in no wise inconsistent with the demands of humanity in cases of emergency. In saying that certain practitioners are not to be considered as regular or fit associates in consultation, it is neither said nor implied that a physician should not see a patient, even with these practitioners, when humanity requires him to do so. The tenor and spirit of the code throughout are opposed to any act of professional inhumanity. Moreover, in particular cases the physician must be the judge of his duty in this regard."

The view of the code here taken is certainly a novel one, and one that, so far as we are aware, has never before been publicly advanced. Certainly the American Medical Association has never given its official sanction to this explanation of its consultation clause, nor has any other society, when called on to enforce the code, accepted such from delinquents as a sufficient excuse for their misdoing. It must therefore be considered as a purely personal view, and as such does honor to its promulgator. Divested of unnecessary verbiage, it simply means that the American code permits consultations with homœopaths in emergencies, and when demanded by the dictates of humanity, and makes the individual practitioner the judge of the necessities and proprieties of the case.

Let us compare this with the consultation clause of the State code, the first sentence of which reads as follows: "Members of the Medical Society of the State of New York, and of societies in affiliation therewith, may meet in consultation legally qualified practitioners." The second sentence is in antithesis to, and an explanation of the first, and reads as follows: "Emergencies may occur in which all restrictions should, in the judgment of the practitioner, yield to the dictates of humanity." It will be noted that this code neither obliges, recommends, or encourages consultations with homœopaths; it simply permits them under circumstances which are specified, and leaves the conscience of the individual physician interested to act as the judge of the necessities of the case. In what respect, we may ask, does this code differ from the American code as interpreted by Dr. Flint? We believe the veriest hair-splitter would find great difficulty in establishing even the minutest difference between the spirit and intent of the one code and the spirit and intent of the other. And yet this difference, whatever it may be, is the nominal cause of the hostile attitude of two important portions of the profession. What reason is there, then, for any further prolongation of the contest? None whatever, so far as the merits of the case are concerned, unless perchance Dr. Flint's interpretation of the code should prove not to be the correct one. There can be no question as to the truth of the assertion that until within a very recent period the American code has almost universally been interpreted as absolutely forbidding mixed consultations, under every and all circumstances, the individual practitioner not being permitted to use the slightest discretion in the matter, except at the risk of professional animadversion. Humanity or emergencies found no place in the bosom of the heresy-hunter, whose special delight, apparently, was to detect some unfortunate practitioner whose heart had gained the better of his prudence. The issue here is plain. Either the old and orthodox interpretation of the code must be accepted, or else the one offered by Dr. Flint. In the latter event it certainly seems to us preferable to accept the phraseology of the State code, the meaning of which is clear and distinct, than to cling to the American, the language of which apparently permits of the most opposite interpretation. Dr. Flint, however, believes that consultations with homœopaths should be forbidden for reasons which we find for the first time stated. He says: "The true ground for refusing fellowship in consultations, as in other respects, is a 'name and an organization distinct from and opposed to the medical profession.'" . . . "It is to be hoped that the body from which the code emanated—namely, the American Medical Association—will adopt such modifications in the phraseology of this section as will place restrictions on consultations, not on the ground of doctrines or forms of belief, but on separation from and avowed antago-nism to the medical profession." . . . "If homœopathic practitioners abandon the organization and the name, provided they have received a regular medical education, there need be no restrictions on consultations other than those belonging to other portions of the code, whatever thera-peutical doctrines they may hold."

It would appear from the foregoing that Dr. Flint's main objection to the homœopaths, from the consultation aspect, is the fact that they have formed medical associations outside those of the regular medical profession, and not in affiliation with them, and that, as a consequence of this, they should be denied professional recognition, and their patients should be denied the advantages of regular advice when such is needed. It does not appear even that emergencies or the calls of humanity would permit an evasion of the rule. In other words, the homœopathic practi-tioners are to be denied recognition, and their patients punished simply because they have established separate organizations. We believe the existence of separate sectarian organizations to be a great evil—one of the greatest that at present afflict the body medical—but we are not disposed to hold up the homœopaths to utter condemnation on account of their exist-ence, when the medical profession itself is mainly the cause of their exist-ence. We have already shown that the homœopaths did not leave the regular societies *voluntarily* and for the purpose of organizing separate societies, but, in fact, were forced out of the established bodies. A recent writer,* commenting on this very point, says:

"But there is, according to Dr. Flint, still a disqualifying cause which should exclude homœopaths from consultations, and this is the assumption

* Dr. Thomas Hun, in "An Ethical Symposium," New York, 1883, pp. 60, 61.

of a name and organization distinct from and opposed to the regular profession. There is undoubtedly force in this objection, but, if we look at the history of the rise and growth of homœopathy in this country, the objection will be weakened, if not invalidated. Surely the doctor is old enough to remember the persistent efforts made in the beginning by the homœopathists, when as yet they had no organization, to be admitted into our county medical societies, or in the case of members of the societies who adopted homœopathy to resist expulsion. The numerous suits unsuccessfully brought before the courts to compel the societies to admit or retain them sufficiently attest that, if they now have a distinct organization, the fault is not on their side. We thrust them out of doors, and now it comes with a bad grace from us to give as a reason for refusing fellowship with them that they are not in our house."

We have no hesitation, therefore, in asserting that Dr. Flint's proposition savors neither of justice nor propriety, and that some better excuse must be devised for excluding homœopaths from consultation when the demands and needs of the sick render such consultations desirable.

While Dr. Flint's commentary on the code was being published in the columns of this journal the supporters of the American code sought to effect an organization of the physicians of this State in opposition to the State society, and with the avowed purpose of resisting any modifications of the code that did not originate with the American Medical Association. A vigorous canvass of the State was made in behalf of this organization, and numerous signatures were obtained to a paper pledging its signers to stand by the American code. This action necessitated the formation of another association, one opposed to the re-enactment of the old code. This latter body forwarded to each member of the regular profession of the State a postal-card bearing on its back the following words: " I, the undersigned, am opposed to the present code of ethics of the American Medical Association, and approve of the use of all honorable means to prevent its re-enactment in the State of New York." The majority of those who signed this declaration and mailed the card back to New York simply attached their signatures and addresses. Many, however, added a few words of comment. All of the cards that were returned to the city came under the eye of the writer, and from them we have copied the following words of comment :

" and all other codes, as thirty years adherence to it has proved its uselessness."—S. F. McF.

" and so is the ———— County Medical Society as a body."—R. L.

" I am strongly in favor of the abolition of all codes, considering any code unnecessary for the guidance of a gentleman, and useless for the restraint of others."—T. C. W.

" I think that the old code was a good one when adopted, but there are reasons why it should now be rejected."—S. P. S.

" I have practiced my profession since 1845, and do not hesitate to say that

I do not recall the time or instance when the old code governed the conduct of men of good sense or repute."—E. V. K.

"I would much rather prefer no code at all. But, being obliged to choose under existing circumstances, I say, emphatically, give me the new code."—M. G. P.

"I heartily indorse this card. Had the same position been taken twenty years ago it would have been better for the people and the profession."—J. R.

"The more liberty we have, the greater amount of good we can do."—L. B.

"I regard the new code as a most unfortunate and unwise substitute for the old code. Abolish the whole thing."—J. M. N. K.

"I have protested against it for eighteen years, and will do all in my power to aid in its overthrow."—C. H. A.

"I believe in each physician practicing medicine according to the dictates of his own conscience."—F. W. C.

"Hope we will succeed. New York can afford to be not represented at the American Medical Association. If we make a bold stand I think we will win —we have the right side, any way. *I am for no code.*"—R. F.

"After obeying and carefully observing the rules of the old code for many long years, I have become strongly opposed to it, and am in full accord with the new code; and am willing to do what little I can to maintain it in the State of New York.

"I see its opposers are working hard, almost moving heaven and earth to bring about its repeal; but I do not think they can succeed. . . . It (the American code) is not in accord with the spirit of the age; it is against the common-sense law of the land, and the best sentiment of all classes of society." —W. B. A.

"I think a code of ethics for the medical profession as unnecessary as a book of etiquette for a true gentleman."—H. F. B.

"An unwritten code is as binding to an honorable, honest man, and a written code, however stringent or liberal, will have no influence over the conduct of any others."—H. A. B.

"I prefer not to be tightened up by any code; shall in the future do as I have done in the past—uphold the dignity and honor of my profession, in my own way, to the best of my judgment."—J. R. B.

"and I am opposed to the pretentious and hypocritical old code of the New York State Medical Society."—L. C.

"The new code does not cause irregulars to rejoice, nor does it encourage a single wrong; the new code is legal; it is the voice of the age. It is progress."—S. J. P.

"Most heartily."—F. W. A.

"In the name of humanity, decency, and liberal progress, Amen!"—H. L.

"I did not approve of the change, but, since it has been done, would not turn back."—C. S. P.

"I would prefer no code, but, if we must have one, let it be liberal."—O. C. F.

"believing *no* code as effective as *any* code."—J. H. F.

"because it can't be lived up to in actual practice. We must meet irregulars, and, if gentlemen, we must treat them as such."—P. K. S.

"I am in favor of free consultations."—J. T. L.

"Patients first, ethics next, and liberal opinions all the time."—C. M. McL.

"I am in favor of the present code of ethics of the American Medical Association, and approve of all honorable means to effect its re-enactment in the State of New York, hoping that we may soon have it amended in a way that will leave all at liberty to *council* with all whom they please, and not admit (as the present code does) that the Legislature is competent to say with whom it is proper and right to meet in consultation. A code of medical ethics should ignore all sects in medicine."—L. B.

The foregoing comments certainly indicate a wide diversity of feeling on the questions involved in the present discussion, and the different standpoints from which they are viewed.

The next important event in connection with the code controversy occurred at a meeting of the New York Academy of Medicine in April last. This body, by virtue of its charter, is entitled to representation in the State society. Its by-laws, however (unlike those of the county societies), are not subject to the revision of the State society. The Academy has generally been regarded as the peculiar stronghold of the conservatives, it being claimed that they possessed a large majority in that body. Some question having arisen as to the right of the Academy to representation in the American Medical Association after the exclusion of the New York State society, it seemed important to the old-code party to place the Academy distinctly on record as a supporter of the American code. At a meeting of the Academy, held on the 19th of April, Dr. Austin Flint, Jr., introduced a series of resolutions disavowing sympathy with the action of the State society, and pledging the Academy to renewed allegiance to the code of the American Medical Association. The Committee on Admissions of the Academy was also directed to report for membership only such persons as would pledge themselves to support the old code. The resolutions were adopted by a large majority, obtained by the very simple expedient of assembling the old-code members by means of a secret circular, and without notice of the proposed action to the other side. The detailed proceedings of this meeting of the Academy have obtained a very wide publicity, and need not here be rehearsed. It was at this meeting that Dr. Flint, Jr., first appeared as the virtual leader of the old-code element, and indicated that the policy to be pursued would be characterized by the *fortiter in re*, rather than the *suaviter in modo*. In other words, opposition was to be overcome by brute force, rather than by an appeal to argument and reason. This, indeed, was so thoroughly characteristic of the methods that for years have prevailed in the American Medical Association itself that we need not be surprised at anything that is done in its name. The effect of this action in the Academy was not all that was hoped by its supporters. Instead of strengthening the old-code party in this city, it distinctly weakened it. Many gentlemen whose bias was in favor of the old code, and who were willing to vote for its restoration, recoiled at the means that the leaders of their party seemed disposed to adopt, and declined to follow them farther. This was most conclusively shown a few months later.

The American Medical Association met early in June, and when the members assembled they found themselves confronted with an order from the Judicial Council, notifying them that they would not be permitted to register and take part in the proceedings of the meeting unless they signed a pledge of fidelity to the code of ethics. This pledge terminated with the words : " I will use my best efforts to maintain the same, and in testimony whereof, I hereunto affix my name." A careful examination of the constitution and by-laws of the association fails to discover any authority for this action of the Judicial Council. It was a pure assumption on their part, for which they possessed no warrant whatever. This action of the council does not surprise us ; in fact, nothing that this body should do, or attempt to do, would surprise us. We were, however, immeasurably surprised that any of the members of the association were willing to be thus deprived of their liberty of action, or to deprive their fellow-members of theirs. We believe that the majority of the signers could hardly have been aware of the full intent of the pledge they were singing, as it virtually binds them to use their best efforts to maintain forever, without change, the present code of the American Medical Association.

We see in this act of the Judicial Council and of the association a striking example of the dangers that beset every unchartered and irresponsible body; a clique, once getting into power, hold the members at their mercy, and are enabled to trample on their rights at any moment, without fear of being held accountable, either as individuals or as an association. Not so in a chartered society, even with by-laws identical with those of the American Medical Association. Such a society would have been compelled to admit its regular members, irrespective of signing or not signing such a pledge. Doubtless the old-code members of the New York Academy of Medicine would be very glad to exclude from the meetings such of its members as do not approve its present code and by-laws ; but they know such an attempt would prove futile. In a chartered society, every member knows, or can readily learn, his rights, and neither a ring, clique, nor even a majority in the society, can deprive him of them against his will. In an unincorporated society, however, anything may be done that at any time a majority approves, and there is no redress for those who may be injured thereby. For instance, in the American Medical Association a majority could, by mere vote, pass a resolution expelling all members who are opposed to them, and the expelled members would not be able to defend, or regain, their rights through an appeal to the courts. In fact, something very like this was done a few years ago. The writer attended a meeting of the association in 1865, and paid his fee of five dollars, which at that time entitled him to life-membership, without further payment of dues. A few years later he, in common with other members who had joined the association on similar terms, was informed that his life-membership would be forfeited unless he maintained it by a further annual payment of five dollars. This was a most unmistakable breach of contract, and violated the commonest principles of honesty and morality. The instance is cited simply as an example of the many arbitrary acts of the association, and to show how any act may be done by any voluntary, unincorporated society. Despite these facts, there are those who desire to see the incorporated societies of this State subjected to the control of such a body. We can hardly believe that the gentlemen holding these views have given any very careful consideration to the subject. In common with the majority of the profession, we approve of the existence of a national medical association, but it should be one devoted to scientific pursuits only, and should not attempt to interfere with medical politics in any manner. If the American Medical Association would reorganize on such a basis, New York State, we believe, would be unanimous in its support. If it continues as it is, its existence is but a question of a very few years.

The refusal of the American Medical Association to permit those who did not approve its code to participate in the proceedings of the Cleveland meeting, thereby preventing any discussion of the subject, has proved to be one of the most valuable allies to the cause of the New York State code. It was generally supposed that the advocates of the American code in this State would endeavor to secure desirable modifications at the Cleveland meeting, and many signatures were obtained to the old-code papers in consequence of this impression. The warrant for this belief was the fact that Dr. Flint in his commentaries had expressed his dissatisfaction with the code, as already shown in the present paper, together with the peculiar wording of the papers sent for signatures. In fact, many signatures were obtained on the express representation that a change of the code would be attempted by the members from New York. No such attempt was made, and some of the old-code signers, now fully appreciating the fact that no change may be expected by the association, and not approving the code as it stands, have withdrawn their support and influence, and have given them to the State code. We believe that a majority of the old-code supporters gave their signatures not from a fondness for the old code, but from a belief that all changes in it should originate with the association. Many have the idea that the association possesses some sort of jurisdiction over the profession of the State, and that resistance to its by-laws is a species of rebellion against constituted authorities. This idea is an absolutely false and mistaken one. The association has no more jurisdiction over the different State societies than the American Gynæcological or the American Dermatological Society has. Moreover, there is no process by which it could obtain such jurisdiction. No single State in the Union could give it a charter that would enable its power to be exerted beyond the limits of the State granting the charter ; while the best that the United States could do would be to grant a charter the jurisdiction of which would be coextensive with the District of Columbia. Congress has, for instance, the power to establish a medical college within the limits of the District, and to make the diplomas of the college licenses to practice within that area, but it does not possess the power to make them valid in any State in the Union. Such power exists solely in the respective States.

The profession of this State, and of other States as well, should remember that they are not doctors by divine right, or the grace of God, but simply through the will of the various State Legislatures. It is this which gives them a legal right to call themselves "doctors," and permits them to practice their profession. The Constitution of the United States guarantees a certain amount of freedom in the exercise of religious privileges, but it makes no such guarantees as regards the exercise of the legal, medical, or any other profession or trade. These rights the States reserved at the time of the formation of the Union, and since then have never yielded their prerogative to the national authorities. The action of the New York State society has been likened to the action of the Southern States at the commencement of the late "unpleasantness." It should be remembered that at the time of the formation of the Union the various States entered into a compact with each other, and that their secession was a breach of that compact. The Medical Society of the State of New York, however, never formed any compact, or entered into any contract with the American Medical Association or with any of the other States, nor, so far as we are aware, did any other State. There is therefore no analogy between the relations of the various States to the Union and the purely voluntary relation of the different State societies to the American Medical Association. The latter can, at any time it chooses, and for any cause, refuse to admit the delegates from any of the States. In like manner any of the State societies can sever its existing connection with the American Medical Association whenever it deems it to be to the interest of the profession of the State to do so.

In 1882 the New York society considered that it was better to relinquish its connection with the American Medical Association than to continue the connection, subject to the objectionable by-laws of the latter body. The association, on the other hand, thought it would be better to dispense with the representatives from New York than to alter its by-laws. This it had a perfect right to do, and no one, so far as we are aware, has found any fault with it for so doing. The only power possessed by the association is of a moral nature; legal power it has none. We should therefore judge its actions by the moral standard alone. The repudiation of its financial contract with its early permanent members, and the recent refusal to admit certain of its members, who under its by-laws were entitled to admission, should be sufficient to place the seal of condemnation upon the association, judged by the standard we have alluded to.

During the summer months there was an apparent cessation of active operations by both parties. Early in October, however, Dr. Fordyce Bar-

ker, the president of the New York Academy of Medicine, sent to each of
the members a recommendation that the by-laws of the Academy be altered
in certain respects. The Academy came into existence by virtue of a charter from the State granted in the year 1847. This charter conveyed certain rights and privileges, none of them, however, of a medico-political nature. The Academy, shortly after its organization, assumed such powers,
and, through its moral influence, exercised them for many years. The
alterations of the by-laws proposed by Dr. Barker involved a repudiation
of its former political aspirations, and the resumption of a purely scientific
status, as contemplated in its original charter. To accomplish this end it
would be necessary for the Academy to *repeal* its allegiance to the American code. This would require a three-fourths vote of its members. When
the matter came to a decision, it was found that a three-fourths vote was
not in favor of the repeal of the code, but, to the surprise of many, there
was a very decided majority in favor of such action. The Academy therefore stands to-day as adverse to the old code, but without power to repeal
it, while the minority who are in favor of the old code have no power to
enforce it against the wishes of a stronger adverse sentiment. This is certainly an anomalous state of affairs, which can not continue for any great
length of time.

For some months it had been claimed by the supporters of the old code
that the general sentiment of the profession in this city was in favor of
the re-enactment of the American code. It was asserted that a majority
of the County society were in favor of such action. The test was made at
the annual election of officers in October. On this occasion both parties
brought to the polls their full voting strength. On counting the votes,
it was found that there were 220 in favor of restoring the old code, and
375 opposed to so doing. It seems quite certain that the old-code association had secured, early in the spring, a sufficient number of supporters to
give them hopes of success. When the matter came to a vote, however,
both in the Academy and in the County society, the result showed that
there had been many breaks from the old-code ranks. Many gentlemen,
who in the spring had hastily signed the old-code papers, on careful
examination of the subject reconsidered their action. Many others regarded the occurrences that took place at the April meeting of the Academy of Medicine a sufficient ground for withdrawing their sympathies
from the leaders of the majority on that occasion. A still greater number, however, we are satisfied, abandoned the fortunes of the old-code
party in consequence of the failure of the leaders of that party to even
attempt to secure changes in the American code that so many of their
followers considered desirable.

We may, I think, consider the code question as definitely settled, in
this city at least, so far as regards the restoration of the code of ethics
of the American Medical Association. The County society, the only
body that can legitimately consider the question, at the last election, as

on every previous occasion on which the matter has been brought forward, distinctly expressed its sentiments in opposition to the old code. It is true that the question can hardly be said to have reached a final solution in the Academy of Medicine. This body, with a limited membership, possesses but a limited influence on the mass of the profession, and it is of very little practical consequence which way it is there decided. The liberals, it is true, are in the majority, and will, without the slightest doubt, remain so, and may expect accessions both from the ranks of the present conservatives and from among those who in the future become members. As we have already shown, the code has been for many years practically in abeyance in the Academy, so far as its enforcement was concerned, and there is but little likelihood that its vitality will ever again be tested in that body. Such being the case, we are perfectly willing that those who profess to admire the code should continue to wear it as an ornament and a phylactery.

Before closing this series of papers, one or two questions of importance require notice. As we have already shown, certain of the county societies of this State repudiated the action of the State society in the matter of the code, and declared that they would stand by the old code and retain it among their by-laws.

It has been thought by some that such action on the part of the county societies would forfeit their title to representation in the State society. Such, however, is not the case. There is no provision in any of the statutes by which the State society is empowered to deny the county societies the right of representation, no matter how rebellious they may be as regards the edicts of the former body. The only way in which the State society can defend itself in the matter is by a direct application to the Legislature to have the charters of the offending societies revoked, as provided for in Sec. 2 of Article V of the by-laws of the State society. There is little doubt that, in the present state of public opinion, if such application were made to the Legislature, the request of the State society would be promptly granted. The county societies, however, should be aware that any code or by-laws they adopt contrary to the wishes of the State society are absolutely illegal and null and void. The importance of this matter suggested to the writer the advisability of obtaining the opinion of legal counsel concerning it, and to this end we submitted the following question : What will be the effect if a county medical society adopts a by-law which is not in accordance with the ordinances of the Medical Society of the State of New York, or which does not receive the approval of said State society ? The answer to this question was as follows :

"By the provisions of Sec. 14 of the Act of 1813 (Chap. 94), it is made 'lawful for the respective' (county) 'societies to make such by-laws . . . . as they shall think fit and proper, provided . . . . that the by-laws . . . . shall not be repugnant to the by-laws, rules, and regulations of the Medical Society of the State of New York.'

"And, by the provisions of Sec. 1 of Chap. 445 of the laws of 1866, it is declared to be 'lawful for any county medical society in this State . . . . to establish such rules and regulations for the government of its members as they may deem fit, provided the action of such societies receive the sanction of the said State medical society.'

"I am, therefore, of opinion that any such action on the part of a county society would be null and void and of no effect; in other words, as the law stands, the county societies can not adopt a by-law such as is suggested by the question, and any attempt to do so would be idle and of no avail."

We here leave the question of the codes and the status of the profession, with the statement that we have endeavored to be accurate as to facts, logical as to inferences, and moderate but candid in the expression of opinion. We can not but hope that the whole matter will receive a speedy solution, and one that will commend itself to the great majority of the profession in this State.

The New York Medical Journal,

A WEEKLY REVIEW OF MEDICINE.

PUBLISHED BY
D. Appleton & Co.

EDITED BY
Frank P. Foster,
M. D.

THE NEW YORK MEDICAL JOURNAL, now in the nineteenth year of its publication, is published every Saturday, each number containing twenty-eight large, double-columned pages of reading-matter. By reason of the condensed form in which the matter is arranged, it contains more reading-matter than any other journal of its class in the United States. It is also more freely illustrated, and its illustrations are generally better executed, than is the case with other weekly journals.

It has a large circulation in all parts of the country, and, since the publishers invariably follow the policy of declining to furnish the JOURNAL to subscribers who fail to remit in due time, its circulation is *bona fide*. It is largely on this account that it is enabled to obtain a high class of contributed articles, for authors know that through its columns they address the better part of the profession; a consideration which has not escaped the notice of advertisers, as shown by its increasing advertising patronage.

The special departments of the JOURNAL are as follows:

LECTURES.—The frequent publication of material of this sort is a prominent feature, and pains are taken to choose such as will prove valuable to the reader.

ORIGINAL COMMUNICATIONS.—In accepting articles of that class, regard is had more particularly to the wants of the general practitioner. and all the special branches of medicine are duly represented.

BOOK NOTICES.—Current publications are noticed in a spirit of fairness, and with the sole view of giving information to the reader.

CLINICAL REPORTS are also a regular feature of the Journal. embracing clinical records from the various hospitals and clinics, not only of New York, but of various other cities. together with clinical contributions from private practice.

EDITORIAL ARTICLES are numerous and carefully written, and we are able to give timely consideration to passing events.

MINOR PARAGRAPHS.—Under this heading are given short comments and notes on passing events.

NEWS ITEMS contain the latest news of interest to the profession.

OBITUARY NOTES announce the deaths which occur in the ranks of the profession, with a brief history of each individual when practicable.

SOCIETY PROCEEDINGS are given promptly, and those of a great number of societies figure. At the same time we select for publication only such as we think profitable to our readers.

REPORTS ON THE PROGRESS OF MEDICINE constitute a feature of the Journal which we have reason to think is highly valued by our readers.

MISCELLANY includes matter of general interest, and space is also given for NEW INVENTIONS and LETTERS TO THE EDITOR.

As a whole, we are warranted in saying that the NEW YORK MEDICAL JOURNAL is regarded with the highest favor by its readers and by its contemporaries.

Subscription price, $5.00 per annum.